Analytic Hierarchy Process - Models, Methods, Concepts, and Applications

Edited by Fabio De Felice and Antonella Petrillo

Published in London, United Kingdom

Analytic Hierarchy Process - Models, Methods, Concepts, and Applications
http://dx.doi.org/10.5772/intechopen.1001519
Edited by Fabio De Felice and Antonella Petrillo

Contributors
Hatice Esen, Vidana Gamage Pavani Lakshika, M.J.M. Fazeela Ahzan, Irit Talmor, Mehrnaz Ramzanpour, Rouhollah Rahimi, Vasili Romanchak, Gheorghe Jurj, Federico de Andreis, Enza Curcio, Federico Maria Sottoriva, Ubaldo Comite, Fatih Sivrikaya, Gonca Ece Özcan, Korhan Enez, Mahya Rostampoor, Mustika Sari, Reni Dian Octaviani, Ana Paula Oening, Débora Cintia Marcilio, Natalie Cristine Scrobot, Danielle de Freitas, Patricio Rodolfo Impinnisi, Cleverson Luiz da Silva Pinto, Kartic Bera, Jatisankar Bandyopadhyay, Pabitra Banik

First published in London, United Kingdom, 2023 by IntechOpen
IntechOpen is the global imprint of INTECHOPEN LIMITED, registered in England and Wales, registration number: 11086078, 5 Princes Gate Court, London, SW7 2QJ, United Kingdom

British Library Cataloguing-in-Publication Data
A catalogue record for this book is available from the British Library

Additional hard and PDF copies can be obtained from orders@intechopen.com

Analytic Hierarchy Process - Models, Methods, Concepts, and Applications
Edited by Fabio De Felice and Antonella Petrillo
p. cm.
Print ISBN 978-1-83769-541-6
Online ISBN 978-1-83769-540-9
eBook (PDF) ISBN 978-1-83769-542-3

Meet the editors

Fabio De Felice is a professor in the Department of Engineering, Parthenope University of Naples, Italy. He received his Ph.D. in Mechanical Engineering from the University of Cassino and Southern Lazio, Italy. His current research focuses on multi-criteria decision-making analysis (with emphasis on Analytic Hierarchy Process [AHP] and Analytic Network Process [ANP]) and industrial, project, and supply chain management. Currently, he serves as a member of the Scientific Advisory Committee of the International Symposium on the Analytic Hierarchy Process (ISAHP). He is the founder of AHP Academy, which promotes the diffusion of the culture and methodologies of decision-making, with particular reference to those based on AHP. He is a member of the editorial boards of several international organizations and journals and has authored/co-authored numerous articles on decision science and business management.

Antonella Petrillo, Ph.D., is a professor in the Department of Engineering, Parthenope University of Naples, Italy. She received her Ph.D. in Mechanical Engineering from University of Cassino and Southern Lazio, Italy. Her research interests include multi-criteria decision analysis, industrial plants, logistics, manufacturing, and safety. She serves as an associate editor for the *International Journal of the Analytic Hierarchy Process*. She is a member of AHP Academy and a member of several editorial boards.

Contents

Preface

Many of us are used to making decisions in a rather linear way. We are used to analyzing the scenario, identifying causes, effects, and options, weighing the advantages and disadvantages, and finally deciding. However, when there are too many options or alternatives, each with comparable advantages and disadvantages, it becomes much more difficult to take the right decision. The real world is far from being linear.

In fact, we live our lives in complex contexts that constantly place us in front of decision-making processes. And this happens in both personal and professional life. Deciding means taking responsibility, giving answers, giving direction, realizing, starting, or concluding.

The book comes from the awareness that there is no "business" without decisions. Definitively, all of us are decision-makers in the professional and personal spheres. Promoting awareness of decision-making mechanisms and applying rigorous methods and techniques is essential for making the right decisions at the right time.

In a complex context of decisions, it is useful to apply accurate and valid scientific methods. There are many scientific decision-making methods, among which the best known is the Analytic Hierarchy Process (AHP).

AHP is a multi-criteria decision support technique developed in the 1970s by an Iraqi mathematician naturalized in the United States named Thomas L. Saaty. The methodology makes it possible to compare several alternatives in relation to a plurality of criteria, of a quantitative or qualitative type, and to obtain a global evaluation for each of them. The strength of the AHP is that it provides for a distinction between the subjective component of the evaluation and the objective data. The decision-maker identifies a set of criteria on the basis of which to evaluate the decision alternatives and assigns a percentage weight to each criterion, after which the decision-maker assigns a score, which is the impact of the criterion on the decision. The score of each decision alternative is the weighted average of the scores of each criterion on the decision by the weight assigned to each criterion. The most relevant characteristics of the AHP are the ability to measure intangible criteria, flexibility, and applicability in solving complex problems.

Based on these premises, this book collects interesting research and studies that highlight the benefits and power of the AHP through analysis of different case studies (practical and methodological) using the process.

In particular, the book proposes several applications of AHP for decision-making in transportation organizations, the medical field, construction, personnel selection, for environmental decisions, and more. All applications proposed demonstrate the importance of using a scientific approach for strategic planning in any kind of decision. There is no better way to approach a complex problem than to structure all the important key influences affecting every important alternative course of action that we can think of. Certainly, decision-making today is indeed a science.

Based on the studies presented, this book is a useful resource for anyone who deals with decision-making with a focus on AHP. Furthermore, we hope that this book will provide useful resources, ideas, techniques, and methods for further research in the field. Special thanks to all the authors who contributed to the success of the project. We thank the authors who accepted our invitation to contribute their invaluable research as well as the referees who reviewed these papers for their effort, time, and invaluable suggestions. Our special thanks to Author Service Manager Dr. Ivana Barac and her team at IntechOpen for their support.

Finally, we want to remember our dear friend Tom Saaty, a special man, father of the AHP, and an unforgettable mentor. We are thankful for the many memories of the Pittsburgh house and the long discussions on AHP. We learned a lot from Tom, especially the importance of "thinking" and carving out the time to think and avoiding lazy habits of not thinking. Thank you for all the moments spent together and for your precious teachings. You will always be in our thoughts.

Fabio De Felice and Antonella Petrillo
Department of Engineering,
University of Naples "Parthenope",
Naples, Italy

Section 1

Theory and Methods

Chapter 1

Analytical Hierarchy Process Problem Solution

Hatice Esen

Abstract

Multi-criteria decision-making methods have been developed to help people or managers in the decision-making process and make the right decision. The Analytical Hierarchy Process (AHP) method is one of the most frequently used multi-criteria decision-making methods and is used in situations where the decision-makers' objective and subjective decisions need to be evaluated together. This study explains the application of the AHP method in personnel selection in Excel. AHP is an easily applicable method for solving complex problems. In this study, the problem of decision-making of nurses who apply to the intensive care nursing certificate program is handled. The AHP method was used in the solution to the problem. The three nurses have been selected by considering the criteria determined by the training officer who organized the certificate program among the nurses who applied to the program by ranking. The use of AHP in addition to the traditional method at every stage of human resources planning will contribute to the personnel selection process and the efficiency of health service delivery.

Keywords: AHP, multi-criteria decision-making, decision, problem solving, decision-making

1. Introduction

Decision-making is one of the most important activities performed by managers and is also known as the core of management. The decision-making process can be defined as the activity that leads to the solution of a decision-making problem involving at least two alternatives, and the chosen one gives the best result according to the determined goal. The decision-making process is carried out by managers intuitively or using decision-making methods [1]. The Analytical Hierarchy Process (AHP) is a frequently used multi-criteria decision-making (MCDM) method, which is a mathematical approach. AHP is a widely used method to provide solutions to multi-objective decision-making problems, developed by Thomas L. Saaty (1970) and was brought to the agenda by Myers and Alpert (1968) [2]. It is a method [3] using the decision-makers experience, knowledge, and intuition [4] to make objective and subjective decisions [5]. In other words, AHP is a method that aims for people to make their own decisions instead of obliging how to make decisions [6].

AHP has a decision hierarchy with some characteristic features applied to various problems. This hierarchy includes decision objectives, criteria, sub-criteria, and alternatives [7]. After determining the hierarchical structure, AHP determines the relative priorities of the alternatives according to the decision-makers preference. At each level of the hierarchy, the decision-maker needs to make two-way comparisons using the alternatives and the importance of the criteria [8, 9].

In addition to its ease of use, AHP is an easy and successful application for solving complex problems that contain objective and subjective judgments [10]. The method is used when there is more than one decision-maker among many conflicting alternatives under certainty and uncertainty [11]. In other words, the AHP methodology allows us to determine which alternative best fits our criteria and the level of importance we attach to them [12].

It is a well-known and widely used multi-criteria decision-making method in various fields, such as machine selection, supplier selection, ambulance allocation, and prioritization of nurses and other resources in health services [6]. For example, for the global forecast of diabetes patients for 2011–2030 [13], to evaluate the performance of hospitals in Taiwan [14], to determine the most appropriate treatment for rectal cancer treatment [15], to compare pregnant women's recommended delivery [16], to determine efficacy in colorectal cancer treatment [17], to determine dental treatment priorities [18], to determine the criteria to be used in the decision to admit COVID-19 patients to intensive care units [19], and to design and build a decision support system application [20].

AHP attempts to determine the relative importance of each element of a decision by comparing each pair of alternatives at each hierarchy level. Thus, the technique examines multiple alternatives and provides the decision-maker with the relative priorities of the various alternatives under consideration. When applying AHP to a real-life problem, a decision-maker can convert his subjective view to an objective one, which gives the decision-maker confidence that their intuition and experience are not overlooked when arriving at the final ranking of alternatives. The technique's simple nature and low mathematical complexity make it the most preferred in various industries, including engineering, healthcare, finance, public policy, and business environments [6].

This study consisted of multi-criteria decision-making AHP, AHP implementation steps, and sample AHP works sections in the introduction of the study. The study consisted of five sections. In the first part, general information about multi-criteria decision-making and AHP is given. In the second part, the strengths and weaknesses of AHP were explained. In the third part, AHP steps are mentioned. In the fourth chapter, the solution to the personnel selection problem with AHP is explained and in the last part, the study was evaluated and recommendations made for future studies.

2. Strengths and weaknesses of the AHP method

The strengths and weaknesses of the AHP method are presented below.

2.1 Strengths and weaknesses of AHP

The strengths of AHP are described below [21–23].

- It makes the content of the problem easier to understand.

- It arranges complex and multi-criteria problems hierarchically.
- Quantitative variables quantify the quantitative and qualitative criteria, allowing them to be evaluated together.
- It evaluates the consistency of the objective and subjective judgments of the decision-maker.
- It provides a flexible model used in complex and large-scale problems.
- The ease of reaching a consensus allows for new insights to emerge, and results are more reliable.
- It has ease of application in almost every field.

2.2 Weaknesses of AHP

The weaknesses of AHP are described below [21–23].

In cases where the number of alternatives and criteria is high, it can be challenging to create pairwise comparison matrices.

- Problems can arise when the criteria are not very clearly defined.
- There is no independent method that validates the results as it is based on and guided by the personal judgments of the decision-makers.
- When criteria are added to the analysis later, the whole process may need to be repeated from the beginning.
- The comparison process may take a long time if the decision-maker is not a single person but a group.

3. Stages of the analytical hierarchy process

In the first stage of AHP, the problem is clearly defined, and alternatives are determined. Then, the hierarchical structure for solving the problem consists of determining the advantages according to the degree of importance [4] and determining the alternatives if the consistency ratio is acceptable [9].

The stages of the problem-solving process with the AHP method are shown in **Figure 1**.

3.1 Creating the hierarchical structure

The first step in AHP analysis is clearly defining the purpose [1] and what is called decision modeling is to create a hierarchy [12]. At the top of the hierarchy, the decision-maker has the ultimate goal to be achieved. With this structure, comparisons can be made easily between the criteria and alternatives determined to achieve the goal [25]. The points to be considered in the design of the hierarchy were explained by Saaty (1990) as follows:

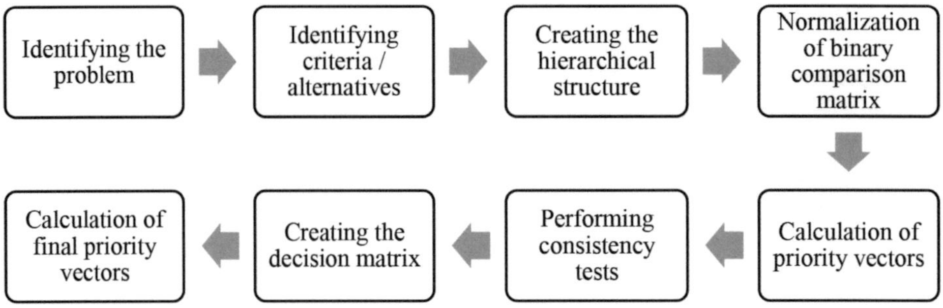

Figure 1.
AHP solution process. Source: Ref. [24].

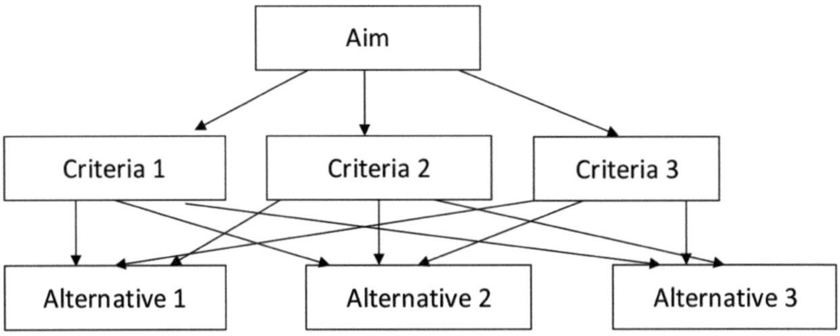

Figure 2.
AHP hierarchical structure. Source: Ref. [5].

- The problem should be explained clearly and precisely.
- The problem should cover all factors.
- All information that will affect the result should be considered [26].

The hierarchical structure of the AHP method is shown in **Figure 2**.

3.2 Creating the binary comparison matrix and determining the weights

After the hierarchical structure is established in AHP, the criteria are compared relative to the importance of determining the priorities [25, 27]. The person or persons who are experts on the subject determine the degree of importance by comparing the criteria at each level in pairs [1]. The comparison matrix for the determining criteria is shown in **Table 1**. In a matrix with n elements, n (n-1)/2 comparisons are made [21]. For example, comparing criteria 2 and 3 should be evaluated as "which of these two criteria is more important and how important is it to achieve the goal" [22]. Since the relevant factor is compared with itself, the diagonal of the matrix takes the value of 1 [21].

	Criteria 1	Criteria 2	Criteria ...	criteria j
Criteria 1	W1/W1	W1/W2	...	W1/Wj
Criteria 2	W2/W1	W1/W1	...	W2/Wj
Criteria ...	...	...	...	...
Criteria j	Wj/W1	Wj/W2	...	Wj/Wj

Table 1.
Pairwise comparison matrix.

Importance	Definition	Explanation
1	Equally important	Compared alternatives contribute equally to the defined criteria.
3	Moderately important	The first factor is moderately important compared to the other to achieve the goal.
5	Quite important	The first factor is quite important compared to the other to achieve the goal.
7	Much more important	Factor 1 is very strongly important over the other.
9	Extremely important	The first factor is extremely important compared to the other.
2,4,6,8	Intermediate values	Used when compromise is needed.
Mutual Values	If the value of "x" is compared with the value of "i" and "j"; j will be (1/x) when comparing with i.	
Source: Refs. [5, 29].		

Table 2.
Comparison scale.

After the hierarchy is established, it is necessary to calculate how important the criteria are to each other, which is their relative importance. The decision-maker determines the degree of importance between the criteria based on the 1–9 scale [28]. The comparison scale developed by Saaty is explained in **Table 2**.

Since not all criteria are of equal importance, relative priorities (weights) are obtained [12]. According to the AHP method, the eigenvalues and eigenvectors of the comparison matrix help determine the priority order. The eigenvector corresponding to the largest eigenvalue determines the priorities. The largest eigenvalue of the matrix A is λ_{enb} taken as the W priority vector; $(A-\lambda_{enb} I)\ W = 0$ is obtained by solving the system of equations [30].

3.3 Calculation of eigenvector

For each element in the pairwise comparison matrix, the column vector is calculated with the help of Eq. 1, and the C matrix is created as in Eq. 2 [31, 32].

$$b_{ij} = \frac{a_{ij}}{\sum_{i=1}^{n} a_{ij}} \tag{1}$$

$$C = \begin{bmatrix} c_{11} & c_{12} & \dots & c_{1n} \\ c_{21} & c_{22} & \dots & c_{2n} \\ . & & & . \\ . & & & . \\ . & & & . \\ c_{n1} & c_{n2} & \dots & c_{nn} \end{bmatrix} \tag{2}$$

The priority vector showing the importance levels of the factors is obtained, as shown in Eq. 3.

$$wi = \frac{\sum_{j=1}^{n} c_{ij}}{\text{n}} \tag{3}$$

3.4 Calculating consistency ratio

In the AHP method, it is checked whether the comparisons are consistent after the establishment of pairwise comparison matrices and synthesis. For this, the inconsistency rate is calculated in each pairwise comparison matrix, which should be 10% [28]. If the consistency rate is above 10%, the decisions are assumed to be inconsistent, and it is necessary to review the decisions to find the cause of the inconsistency and correct it [12, 31]. If the degree of consistency is acceptable, the process is continued, and the steps are completed. The steps to be followed in calculating the compliance ratio are as follows [33].

- The weighted sums of the elements in the columns are calculated for each row of the comparison matrix.
- The normal matrix is obtained by dividing the element in each row of the comparison matrix by the total column weight obtained.
- A vector of priorities is formed by averaging each row of the normal matrix.
- Column vector D is obtained from the matrix multiplication of the comparison matrix A and the priority vector w.

$$D = \begin{bmatrix} a_{11} & a_{12} & \dots & a_{1n} \\ a_{21} & a_{22} & \dots & a_{2n} \\ . & & & . \\ . & & & . \\ . & & & . \\ a_{n1} & a_{n2} & \dots & a_{nn} \end{bmatrix} x \begin{bmatrix} w_1 \\ w_2 \\ . \\ . \\ . \\ w_n \end{bmatrix} \tag{4}$$

In order to determine the consistent rate of the evaluations made, the compliance rate (consistency) ratio = CR) is calculated. The following formulas are used to calculate this ratio.

$$\text{CI} = \text{Concordance Index}$$

N	one	2	3	4	5	6	7	8	9	10
Random Index (RI)	0.00	0.00	0.58	0.90	1.12	1.24	1.32	1.41	1.45	1.49

Table 3.
Random value index.

$$\mathrm{RI} = \text{Random Value Index}$$

λ_{max} = It is the average of the values obtained by dividing the elements of the all priorities matrix by the priorities vector [34].

$$CI = \frac{\lambda - n}{n - 1} \tag{5}$$

$$\mathrm{CR} = \mathrm{CI}/\mathrm{RI}. \tag{6}$$

A hierarchical structure is created if the compliance rate is within acceptable values. Random index (RI) is the mean value of randomly derived pairwise comparison matrices based on n number [21, 35]. The random value indices determined according to the number of criteria are shown in **Table 3** [7].

3.5 Determining the ranking of alternatives

At this stage, alternatives are determined following the purpose. Obtained alternative evaluations express the importance of subjective perceptions in the decision-makers alternative selection. The sum of the values of the alternatives is 1, and the most suitable alternative with the highest numerical value is [21].

4. AHP application in personnel selection

Human resource management and personnel selection are very important in health service delivery. Traditional methods or managerial insights are used in determining the clinics where nurses will be employed in public hospitals and/or the nurses who will participate in certification training. One of the most important resources in hospitals is the personnel and personnel selection is one of the situation in which multi-criteria decision-making methods are most used. In particular, the qualifications of health service personnel may increase service quality, patient satisfaction, and quality of care.

In this study, the solution to the problem by using the AHP method with a scientific approach as well as the traditional method in the selection of nurses to participate in a training and research hospital intensive care certificate program is discussed. Thus, the applicability of multi-criteria decision-making methods in hospitals and the selection of appropriate personnel according to the determined criteria were provided. After the problem was defined, the criteria were determined by the program manager, well-experienced and have been running the program for many years, and is a training nurse working in the training unit. For this study, only the decisions of the program manager were included and the opinion of another expert was not sought. Before the AHP method, personnel selection was generated on limited parameters such as education status and working status in the intensive care unit. By using this method, the selection of personnel who can provide the best contribution to the institution would be made.

4.1 Step-by-step procedure of AHP

In the case study, we will analyze a personnel selection example with six criteria and six alternatives in Excel. With the application, we will choose three nurses among six, for the intensive care certification program. The nurse criteria to be selected for the certificate program were determined together with the course supervisor (**Figure 3**). The determined criteria are shown in **Figure 4**.

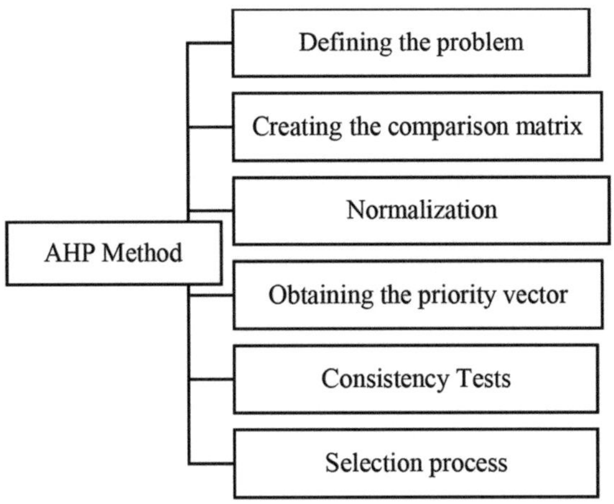

Figure 3.
AHP implementation steps. Source: Ref. [18].

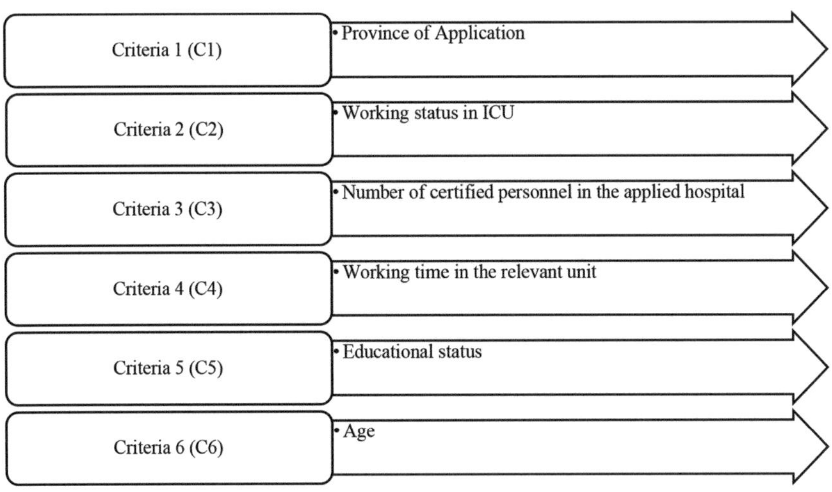

Figure 4.
Criteria for a nurse to be selected for the certificate program.

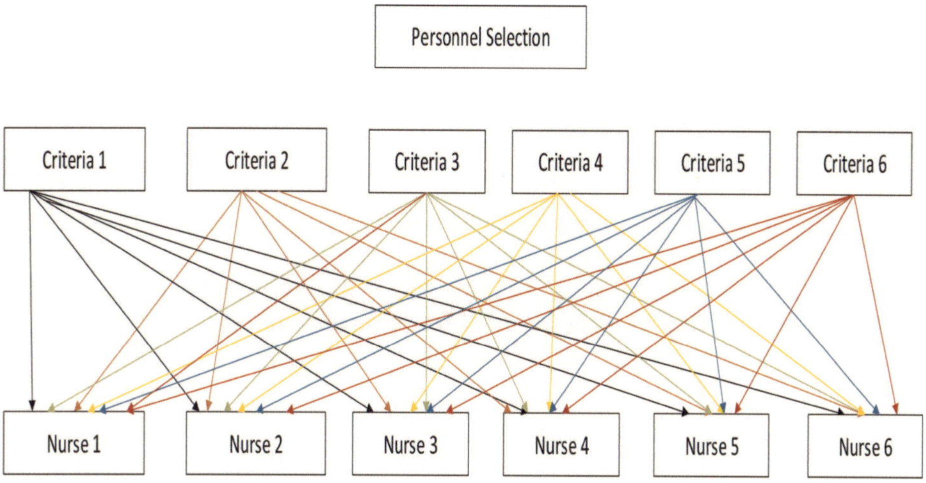

Figure 5.
Personnel selection hierarchy structure.

Personnel selection is among the problems in which multi-criteria decision-making methods are applied. Personnel selection criteria were determined by the program manager by taking into account the requirements of the certificate program guide and the needs of the institution.

The personnel selection hierarchy has been established (**Figure 5**).

In the AHP method, the criteria are first compared among themselves. Then, considering each criterion, the alternatives are compared. The comparison of the criteria is as follows. Since C1 criteria will not have superiority over C1 criteria, this will be 1. This is why the vertices are made up of 1. When we compare the C1 criteria with the C2 criteria, it is determined how important it is according to the degree of importance table. If the C1 advantage over C2 is 9, which is more important, then the value of C2 over C1 in the same matrix will be the opposite, 1/9.

If opinions are taken from experts in more than one field, a comparison of the binary matrix is taken from each of them separately. Opinions from all experts are collected in a single matrix by taking the geometric mean. The comparison matrix is an $n \times n$ square matrix. The matrix components on the diagonal of this matrix take the value 1. In the first stage, the importance level of the pairwise comparison matrix was determined by considering the formulation below (**Table 4**).

	C1	C2	C3	C4	C5	C6
C1	**1**	3	5	2	6	5
C2	1/3	**1**	2	3	3	5
C3	1/5	½	**1**	2	4	5
C4	1/2	1/3	1/2	**1**	3	3
C5	1/6	1/3	1/4	1/3	**1**	3
C6	1/5	1/5	1/5	1/3	1/3	**1**

Table 4.
Determining the significance of the pairwise comparison matrix.

$$A = \begin{bmatrix} a_{11} & a_{12} & \dots & a_{1n} \\ a_{21} & a_{22} & \dots & a_{2n} \\ . & & & . \\ . & & & . \\ . & & & . \\ a_{n1} & a_{n2} & \dots & a_{nn} \end{bmatrix}$$

The normalization of the pairwise comparison matrix obtained in the second step was made. The following formula was used for normalization.

$$b_{ij} = \frac{a_{ij}}{\sum_{i=1}^{n} a_{ij}}. \tag{7}$$

For the normalization process, the sum of each column is calculated. The elements of each column are obtained by dividing the column-by-column sum (**Tables** 5 and **6**).

This matrix was also the normalized matrix.

	C1	C2	C3	C4	C5	C6
C1	1/2.40	3 / (5.37)	5/ (8.95)	2 / (8.67)	6 / (17.33)	5/ (22.00)
C2	(1/3) /(2,40)	1 / (5.37)	2 /(8.95)	3 /(8.67)	3 / (17.33)	5/ (22.00)
C3	(1/5)/ (2.40)	(1/2) / (5.37)	1 / (8.95)	2 / (8.67)	4 / (17.33)	5/ (22.00)
C4	(1/2) /(2.40)	(1/3) / (5.37)	(1/2) / (8.95)	1 / (8.67)	3 / (17.33)	3 / (22.00)
C5	(1/6) /(2.40)	(1/3) / (5.37)	(1/3) /(8.95)	(1/3) / (8.67)	1 / (17.33)	3 / (22.00)
C6	(1/5) / (2.40)	(1/5) / (5.37)	(1/3) / (8.95)	(1/3) / (8.67)	(1/3) / (17.33)	1 / (22.00)
TOTAL	2.40	5.37	8.95	(8.67) / (8.67)	17.33	22.00

Total value refers to the total of each column.

Table 5.
Normalization steps.

	C1	C2	C3	C4	C5	C6
C1	0.417	0.559	0.559	0.231	0.346	0.227
C2	0.139	0.186	0.223	0.346	0.173	0.227
C3	0.083	0.093	0.112	0.231	0.231	0.227
C4	0.208	0.062	0.056	0.115	0.173	0.136
C5	0.069	0.062	0.028	0.038	0.058	0.136
C6	0.083	0.037	0.022	0.038	0.019	0.045

Table 6.
Normalization matrix.

$$C = \begin{bmatrix} c_{11} & c_{12} & \dots & c_{1n} \\ c_{21} & c_{22} & \dots & c_{2n} \\ . & & & . \\ . & & & . \\ . & & & . \\ c_{n1} & c_{n2} & \dots & c_{nn} \end{bmatrix} \tag{8}$$

In calculating the priority vector, the arithmetic average of each row is taken. The formula is shown below (**Table 7**).

$$wi = \frac{\sum_{j=1}^{n} c_{ij}}{\text{n}} \tag{9}$$

The data obtained in the priority vector is also used as weights in the AHP matrix. After this stage, it is checked whether the answers are consistent.

The necessary parameters to perform the consistency tests are shown below.

- Compliance Index (CI)
- Random Value Index (RI)
- Compliance Rate (CR)

For CI we need lambda value. The lambda value is calculated according to the formula below. In other words, the lambda value is obtained by multiplying the priority vector (w) that we have obtained with the columns of the first pairwise comparison matrix. At the same time, vector D is obtained.

$$D = \begin{bmatrix} a_{11} & a_{12} & \dots & a_{1n} \\ a_{21} & a_{22} & \dots & a_{2n} \\ . & & & . \\ . & & & . \\ . & & & . \\ a_{n1} & a_{n2} & \dots & a_{nn} \end{bmatrix} x \begin{bmatrix} w_1 \\ w_2 \\ . \\ . \\ . \\ w_n \end{bmatrix} \tag{10}$$

	C1	**C2**	**C3**	**C4**	**C5**	**C6**	**Priority vector** (w)
C1	0.417	0.559	0.559	0.231	0.346	0.227	0.390
C2	0.139	0.186	0.223	0.346	0.173	0.227	0.216
C3	0.083	0.093	0.112	0.231	0.231	0.227	0.163
C4	0.208	0.062	0.056	0.115	0.173	0.136	0.125
C5	0.069	0.062	0.028	0.038	0.058	0.136	0.065
C6	0.083	0.037	0.022	0.038	0.019	0.045	0.041

Table 7.
Priority vector.

After obtaining the D column vector, the lambda value is obtained with the following formula. Each row of the binary comparison matrix is multiplied by the priority vector (**Table 8**). (= DMULT formula is used).

In order to perform consistency tests;

Compliance Index (CI).

Random Index (RI).

Compliance Ratio (CR) values are required.

Lambda (λ) values are calculated by taking the arithmetic average of the values resulting from multiplying the comparison matrices by their weights and dividing the matrices again by their weight values. The EI value is calculated by dividing the D column vector/priority vector.

Priority vector	D columns vector	EI values
0.390	2699	6925
0.216	1448	6708
0.163	1066	6543
0.125	0.792	6330
0.065	0.408	6241
0.041	0.258	6296

Six values are averaged to calculate the lambda value (λ_{max}).

$$\lambda_{max} = (6925 + 6708 + 6543 + 6330 + 6241 + 6296)/6 = 6507 \quad (11)$$

The lambda (λ) value is subtracted from the number of matrix dimensions and divided by one less than the number of matrix dimensions, and the consistency index is calculated. The consistency ratio is calculated by dividing the consistency index by the randomness value according to the number of matrix dimensions.

$$CI = \frac{\lambda - n}{n - 1} \quad (12)$$

$$\text{CI} = (6.507\text{–}6)/(6\text{–}1) = 0.101. \quad (13)$$

Now, we can calculate the CR value. The required RI value is given in **Table 3**. Since the number of decision alternatives is six, the RI value is 1.24.

	C1	C2	C3	C4	C5	C6	Priority vector	D columns vector
C1	0.417	0.559	0.559	0.231	0.346	0.227	0.390	2699
C2	0.139	0.186	0.223	0.346	0.173	0.227	0.216	1448
C3	0.083	0.093	0.112	0.231	0.231	0.227	0.163	1066
C4	0.208	0.062	0.056	0.115	0.173	0.136	0.125	0.792
C5	0.069	0.062	0.028	0.038	0.058	0.136	0.065	0.408
C6	0.083	0.037	0.022	0.038	0.019	0.045	0.041	0.258

Table 8.
D column vector.

$$CR = CI/RI \tag{14}$$

$$CR = (0.101/(1.24) = 0.08\ (8\%) \tag{15}$$

The result is within the compliance rate limits, below 10%, so the inconsistency is acceptable.

4.2 Identifying alternatives

All alternatives are compared for each criterion. The value of comparing alternatives with each other is 1. Experts are asked to compare the alternatives, and the final value is obtained by taking the geometric mean. After comparing all the alternatives for the criteria, the priority vectors are placed in the matrix in which the alternatives and criteria are compared. The final AHP matrix is then obtained (**Table 9**).

Then, the weight values are multiplied by the value in each row. Then, the row values are summed to get the total value of the alternative. After this step, we can perform the sorting. The ranking is done in Excel with the = rank formula (**Table 10**).

According to the results, the first three nurses with the highest score will be accepted to the intensive care certification program.

Weights	**0.4199**	**0.1905**	**0.1662**	**0.1097**	**0.0716**	**0.0421**
Alternatives	C1	C2	C3	C4	C5	C6
Nurse 1	0.352	0.327	0.402	0.469	0.613	2909
Nurse 2	0.248	0.238	0.261	0.258	0.418	3024
Nurse 3	0.159	0.172	0.181	0.284	0.323	1909
Nurse 4	0.093	0.108	0.120	0.156	0.200	0.580
Nurse 5	0.084	0.089	0.098	0.090	0.085	0.602
Nurse 6	0.064	0.066	0.049	0.046	0.041	0.261

Table 9.
Final AHP matrix.

Weights	**0.4199**	**0.1905**	**0.1662**	**0.1097**	**0.0716**	**0.0421**			**Decision**
Alternatives	**C1**	**C2**	**C3**	**C4**	**C5**	**C6**	**Total**	**Rank**	
Nurse1	0.352	0.062	0.067	0.469	0.044	0.122	1117	1	Selected
Nurse2	0.248	0.045	0.043	0.258	0.030	0.127	0.752	2	Selected
Nurse3	0.159	0.033	0.030	0.284	0.023	0.080	0.609	3	Selected
Nurse4	0.093	0.021	0.020	0.156	0.014	0.024	0.328	4	Not Selected
Nurse5	0.084	0.017	0.016	0.090	0.006	0.025	0.239	5	Not Selected
Nurse6	0.064	0.013	0.008	0.046	0.003	0.011	0.145	6	Not Selected

Table 10.
Personnel selection and ranking.

5. Conclusion

As a result, AHP is a very simple, easy-to-use, multi-criteria decision-making method widely used in solving complex probes, and it can be easily applied in the Excel program without the need for any program.

In this study, the problem of nurse selection for an intensive care certificate program in a hospital, the criteria affecting the selection, and which of the criteria are more important in the selection of nurses were determined. According to the results of this research study, the C1 criterion is more important than the other criteria. In the final part, the top three candidates with the highest scores were accepted to the certificate program.

Personnel selection is a process that needs to be done carefully and consists of some steps. In this process, the criteria to be sought by the personnel to be selected should be determined in advance. The degree of importance of the criteria and the characteristics of the job applicants may differ from each other. In this case, multi-criteria decision-making techniques that allow those who decide on personnel selection to make an objective evaluation play an important role. AHP offers the opportunity to evaluate quantitative and qualitative criteria simultaneously. This method is a useful tool, especially when it is necessary to decide on the selection of more than one candidate. In addition to AHP, it is recommended to use other multi-criteria decision-making methods such as Promethee in personnel selection and in all processes that need to be decided (such as material procurement and purchasing).

Conflict of interest

None.

A. Appendix

Binary comparison matrix.

	A	B	C	D	E	F	G
1		C1	C2	C3	C4	C5	C6
2	C1	1	3	5	2	6	5
3	C2		1	2	3	3	5
4	C3			1	2	4	5
5	C4				1	3	3
6	C5					1	3
7	C6						1
8							

After completing the entry in Excel, the dark-marked areas are obtained by dividing the elements by 1.

	A	B	C	D	E	F	G
1		C1	C2	C3	C4	C5	C6
2	C1	1	3	5	2	6	5
3	C2	0,333	1	2	3	3	5
4	C3	0,200	0,500	1	2	4	5
5	C4	0,500	0,333	0,500	1	3	3
6	C5	0,167	0,333	0,250	0,333	1	3
7	C6	0,200	0,200	0,200	0,333	0,333	1
8		1 / C2					
9		1/ D2					
10		1/ E2					
11		1/F2					
12		1/ G2					
13							

Normalization process, column totals are calculated.

ORTALA... =TOTAL(B2;B7)

	A	B	C	D	E	F	G	H
1		C1	C2	C3	C4	C5	C6	
2	C1	1	3	5	2	6	5	
3	C2	0,333	1	2	3	3	5	
4	C3	0,200	0,500	1	2	4	5	
5	C4	0,500	0,333	0,500	1	3	3	
6	C5	0,167	0,333	0,250	0,333	1	3	
7	C6	0,200	0,200	0,200	0,333	0,333	1	
8	TOTAL	2,400	5,367	8,950	8,667	17,333	22,000	
9								
10								
11		=TOTAL(B2;B7)						
12								

After calculating the column totals, each column element is divided by the corresponding column total. The F4 key is used for fixing in Excel (the $ sign that appears in the formula). With this process, the normalization process is done.

Pano | Yazı Tıpı | Hızalama

B8 =B2/$B$8

	A	B	C	D	E	F	G
1		C1	C2	C3	C4	C5	C6
2	C1	1	3	5	2	6	5
3	C2	0,333	1	2	3	3	5
4	C3	0,200	0,500	1	2	4	5
5	C4	0,500	0,333	0,500	1	3	3
6	C5	0,167	0,333	0,250	0,333	1	3
7	C6	0,200	0,200	0,200	0,333	0,333	1
8	TOTAL	2,400	5,367	8,950	8,667	17,333	22,000
9							
10							
11		=B2/$B$8					
12							
13							

It is calculated by taking the average of the elements in the row to determine the priority vector. Priority vector values are also weight (w) values.

Pano | Yazı Tipi | Hizalama | Sayı

B2 =average(B2:G2

	A	B	C	D	E	F	G	H	I	J
1		C1	C2	C3	C4	C5	C6	priority vector		
2	C1	0,42	0,56	0,56	0,23	0,35	0,23	0,39	→	=average(B2:G2
3	C2	0,14	0,19	0,22	0,35	0,17	0,23	0,22		
4	C3	0,08	0,09	0,11	0,23	0,23	0,23	0,16		
5	C4	0,21	0,06	0,06	0,12	0,17	0,14	0,13		
6	C5	0,07	0,06	0,03	0,04	0,06	0,14	0,07		
7	C6	0,08	0,04	0,02	0,04	0,02	0,05	0,04		
8										
9										

For the discrepancy operation, the swarm vector d and the EI values (for the lambda value) are calculated.

I25 =DÇARP(B14:G14;$H$25:H30)

	A	B	C	D	E	F	G	H	I
23									
24		C1	C2	C3	C4	C5	C6	**priority vector**	D Column Vector
25	C1	**0,417**	**0,559**	**0,559**	**0,231**	**0,346**	**0,227**	0,390	2,699
26	C2	**0,139**	**0,186**	**0,223**	**0,346**	**0,173**	**0,227**	0,216	1,448
27	C3	**0,083**	**0,093**	**0,112**	**0,231**	**0,231**	**0,227**	0,163	1,066
28	C4	**0,208**	**0,062**	**0,056**	**0,115**	**0,173**	**0,136**	0,125	0,792
29	C5	**0,069**	**0,062**	**0,028**	**0,038**	**0,058**	**0,136**	0,065	0,408
30	C6	**0,083**	**0,037**	**0,022**	**0,038**	**0,019**	**0,045**	0,041	0,258

Each row of the binary comparison matrix is multiplied by the priority vector. (=Dmult).

Pano | Yazı Tipi | Hizalama | Sayı

J25 =I25/H25

	A	B	C	D	E	F	G	H	I	J
23										
24		C1	C2	C3	C4	C5	C6	**priority vector**	D Column Vector	Ei values
25	C1	**0,417**	**0,559**	**0,559**	**0,231**	**0,346**	**0,227**	0,390	2,699	6,925
26	C2	**0,139**	**0,186**	**0,223**	**0,346**	**0,173**	**0,227**	0,216	1,448	6,708
27	C3	**0,083**	**0,093**	**0,112**	**0,231**	**0,231**	**0,227**	0,163	1,066	6,543
28	C4	**0,208**	**0,062**	**0,056**	**0,115**	**0,173**	**0,136**	0,125	0,792	6,330
29	C5	**0,069**	**0,062**	**0,028**	**0,038**	**0,058**	**0,136**	0,065	0,408	6,241
30	C6	**0,083**	**0,037**	**0,022**	**0,038**	**0,019**	**0,045**	0,041	0,258	6,296

λ_{max} = (6925 + 6708 + 6543 + 6330 + 6241 + 6296) /6 = 6507

$$CI = \frac{\lambda - n}{n - 1}$$

CI = (6.507–6)/ (6–1) = 0.101.
CR = CI/RI.
CR = (0.101/(1.24) = **0.08 (8% ≤ 10%- acceptable value).**

DOI: http://dx.doi.org/10.5772/intechopen.1001072

Author details

Hatice Esen
Antalya Training and Research Hospital, Antalya, Turkey

*Address all correspondence to: hatice.esen@gmail.com

References

[1] Siekelova A, Podhorska I, Imppola JJ. Analytic hierarchy process in multiple-criteria decision-making: A model example. In: proceedings of International Conference on Entrepreneurial Competencies in aChanging World (ECCW 2020). SHS Web of Conferences 90 ECCW 2020. 2021. p. 01019. DOI: 10.1051/shsconf/20219001019

[2] Feng YJ, Lu H, Bi K. An AHP/DEA method for measurement of the efficiency of R& D management activities in universities. International Transactions in Operational Research. 2004;**11**(2):181-191

[3] Badri MA. Combining the analytic hierarchy process and goal programming for global facility location-allocation problem. International Journal of Production Economics. 1999;**62**(3): 237-248

[4] Vargas GL. An overview of the analytical hierarchy process and its applications. European Journal of Operational Research. 1990;**48**:2-4

[5] Saaty TL. How to make decision: The Analystical hierarchy process. European Journal of Operation Research. 1990;**48**: 9-26

[6] Thakkar JJ. Analytic hierarchy process (AHP). In: Kacprzyk J, editor. Multi-Criteria Decision Making. Vol. 336. Singapore: Springer; 2021. DOI: 10.1007/978-981-33-4745-8_3

[7] Gul M, Guneri AF. Hospital location selection: A systematic literature review on methodologies and applications. Hindawi Mathematics Problems in Engineering. 2021;**2021**:6682958. DOI: 10.1155/2021/6682958

[8] Lee SK, Mogi G, Shin SC, Kim JW. An AHP/DEA hybrid model for measuring the relative efficiency of energy efficiency technologies. In: IEEM 2007 2007 IEEE Int. Conf. Ind. Eng. Eng. Manag. Singapore: IEEE; 2007. pp. 55-59. DOI: 10.1109/IEEM.2007.4419150

[9] An SH, Kim GH, Gwang K, K-In. A case-based reasoning cost estimating model using experience by analytic hierarchy process. Building and Environment. 2007;**42**:2573-2579. DOI: 10.1016/j.buildenv.2006.06.007

[10] Onder G, Onder E. Multi-criteria decision making methods. In: Fatih B, editor. Yıldırım, Emrah Önder. Bursa: Dora Basım 3. Press; 2018

[11] Saat M. An approach to multi-criteria decision making: Analytical hierarchy method. Journal of Gazi University Faculty of Economics and Administrative Sciences. 2000;**2**:149-162

[12] Mu E, Pereyra-Rojas M. Practical Decision Making Using Super Decisions v3. USA: Springer; 2018

[13] Whiting DR, Guariguata L, Weil C, Shaw J. IDF diabetes atlas: Global estimates of the prevalence of diabetes for 2011 and 2030. Diabetes Research and Clinical Practice. 2011;**94**(3): 311-321. DOI: 10.1016/j.diabres.2011.10.029

[14] Tsai HY, Chang CW, Lin HL. Fuzzy hierarchy sensitive with Delphi method to evaluate hospital organization performance. Expert Systems with Applications. 2010;**37**(8):5533-5541. DOI: 10.1016/j.eswa.2010.02.099

[15] Suner A, Çelikoğlu CC, Dicle O, Sökmen S. Sequential decision tree using

the analytic hierarchy process for decision support in rectal cancer. Artificial Intelligence in Medicine. 2012; **56**(1):59-68. DOI: 10.1016/j.artmed. 2012.05.003

[16] Sharma PS, Eden KB, Guise JM, Jimison HB, Dolan JG. Subjective risk vs. objective risk can lead to different post-cesarean birth decisions based on multiattribute modeling. Journal of Clinical Epidemiology. 2011;**64**(1): 67-78. DOI: 10.1016/j.jclinepi.2010. 02.011

[17] Wu CL, Ke TW, Meen TH. Evaluation of intensified colorectal cancer treatment using model based on Delphi method, fuzzy logic, and analytical hierarchy process (DFAHP). Sensors and Materials. 2021;**33**(10): 3499-3512. DOI: 10.18494/ SAM.2021.3538

[18] Khan AMR, Prasad PN, Rajamanoharane SW. Service quality performance measurement Management in Corporate Hospitals Using Analytical Hierarchy Process. International Journal of Manufacturing Technology and Management. 2012;**26**(1/2/3/4):196-212. DOI: 10.1504/IJMTM.2012.051434

[19] Ozkan B, Özceylan EE, Kabak M, Dikmen AU. Evaluation of criteria and COVID-19 patients for intensive care unit admission in the era of pandemic: A multi-criteria decision making approach. Computer Methods and Programs in Biomedicine. 2021;**209**:106348. DOI: 10.1016/j.cmpb.2021.106348

[20] Harjanto S, Setiyowati S, Vulandari RT, Surakarta SN. Application of analytic hierarchy process and weighted product methods in determining the best employees. Indonesia Journal of Applied Statistics. 2021;**4**(2):103-112

[21] Subasi H. Comparison of TOPSIS and AHP Methods Used in Multi Criteria Decision Process and An Application [Thesis]. Istanbul: Marmara University, Institute of Social Sciences, Department of Business Administration; 2011

[22] Aktaş R, Doğanay MM, Gökmen Y, Gazibey Y, Türen U. Numerical Decision Making Methods. 1st ed. İstanbul: Beta Edition; 2015

[23] Timor M. Analytical Hierarchy Process. İstanbul: Türkmenevi Bookstore; 2011

[24] Celikbilek Y. Multi-Criteria Decision Making Methods with Annotated and Comparative Health Sciences Applications. 1st ed. Ankara: Nobel Academic Publishing; 2018. DOI: 10.11611/JMER214

[25] Omurbek A, Simsek N. Selection of online shopping site based on analytic hierarchy process and analytic network process methods. Journal of Management and economics research. 2014;**22**:306-327

[26] Kazancoglu Y. Performing Supplier Selection and Performance Evaluation with Operations Research Techniques in the Logistics Management Process: AHP and DEA [Thesis]. İzmir: Ege University Institute of Social Sciences; 2008

[27] Kargı AVS. Analytical Hierarchy Process Model in the Subcontractor Selection Decision of Managers. Bursa: Ekin Printing Home; 2015

[28] Saaty TL. Deriving the AHP 1–9 Scale from First Principle. Switzerland: Berne; 2001

[29] Young KD, Younos T, Dymond RL, Kibler DF, Lee DH. Application of the analytic hierarchy process for selecting and modeling Stormwater best

management practices. Journal of Contemporary Water Research and Education. 2010;**146**(1):50-63. DOI: 10.1111/j.1936-704X.2010.00391.x

[30] Dagdeviren M, Akay D, Kurt M. Analytical hierarchy process for job evaluation and application. Journal of the Faculty of Engineering and Architecture of Gazi University. 2004;**19**(2):131-138

[31] Tas C, Bedir N, Eren T, Alagas HM, Cetin S. Policlinic evaluation with integrating AHP-TOPSIS methods: An Applicatıon in Ankara. Journal of Health Management. 2018;**2**(1):1-17

[32] Kayhan G. Generation a Hybrid System Using Fuzzy Ahp/Fuzzy Topsis in Human Resource Performance Evaluatıon and an Applicatıon [Thesis]. Turkiye: Erciyes University, Graduate School of Natural and Applied Sciences; 2010

[33] Sevinc T, Eren A. Sorting the KOSGEB support models for SMEs by multicriteria decision making methods. International Journal of Engineering Research and Development. 2019;**11**(1): 409-425

[34] Wu CR, Lin CT, Chen HC. Optimal selection of location for Taiwanese nursing home to ensure a competitive advantage by using the analytic hierarchy process. Building and Environment. ScienceDirect Elsevier. 2007;**42**:1431-1444. DOI: 10.1016/j.buildenv.2005.12.016

[35] Paksoy S. Current Approaches to Multi-Criteria Decision Making. Adana: Karahan Bookstore; 2017

Chapter 2

Association of Nostalgia and Neophilia with Millennial Decision-Making Styles: An Analytical Hierarchy Process (AHP) Approach

Vidana Gamage Pavani Lakshika and M.J.M. Fazeela Ahzan

Abstract

Despite the significant amount of research on consumer decision-making styles (CDMS), there remains a dearth of knowledge on the influence of various factors on these styles. Thus, exploring emerging and understudied variables, such as nostalgia and neophilia, is imperative alongside CDMS. Moreover, in the contemporary era of marketing, millennials have emerged as a key area of interest in research. Therefore, this study aims to contribute to a more rigorous understanding of CDMS among Sri Lankan millennials and the association between nostalgia and neophilia with these styles. The study was conducted using a positivist research paradigm with a deductive approach and a survey strategy, with a sample size of 385 participants. Factor analysis was utilized to identify the types of CDMS among millennial consumers in Sri Lanka, and the analytical hierarchical process was employed to prioritize the CDMS types. Pearson's correlation coefficient was used to examine the correlation between nostalgia and neophilia and each dimension of CDMS. The study's findings revealed the presence of seven CDMS types, with perfectionist decision-making and brand consciousness ranking as the top two. Furthermore, the study discovered a significant correlation between neophilia and nostalgia with CDMS.

Keywords: decision-making styles, millennial, nostalgia, Neophilia, AHP

1. Introduction

Marketing experts need to understand and forecast how different sorts of consumers act while purchasing, consuming, and disposing of various products and services in today's dynamic and ever-changing world to meet their needs [1]. On the other hand, an understanding of consumer behavior can direct to enhanced marketing strategies in the element of firms and organizations, and in addition, it helps and guides to improve public policy. As a result, academics have shown an increasing interest in researching consumer behavior, which encompasses an infinite and diversified topics such as consumer decision-making, internal influences, and external effects. Among these areas, consumer decision-making styles have emerged as one of the most prominent and intriguing areas of research among marketing and behavioral sciences scholars [2, 3]. Thus, it is important for determining customer behaviors, segmenting the market, and providing a signal to marketers about the performance of their marketing strategies in the modern marketing era [2, 3].

Similarly, in the modern marketing period, Millennials or Generation Y have emerged as one of the most intriguing scopes of study [4] as Generation Y is one of the most prominent demographic and age cohorts for marketers to target. Furthermore, they are the largest generation in history [5], and they are entering their prime spending years, with enormous indirect spending power due to their strong influence over their parents [5]. Not only do millennials have a greater effect on their parents' decisions, but they also have a greater influence on other people's purchases and decisions than previous generations [4].

Moreover, millennials are the first generation to have grown up in a digital environment, and they are known as "Digital Natives" because they actively participate in social media platform activities such as sharing, searching for information to consume, and working and playing [5]. Despite the enthusiasm, a number of issues have arisen within this prospective marketing segment [1]. As a result, marketers need to understand and prioritize the decision-making styles of millennials, as well as create innovative strategies to appeal to and overcome the problems associated with this generation of customers. Furthermore, despite the fact that consumer decision-making styles have been studied over the last few decades using a variety of nationalities, including Americans, Koreans, Chinese, New Zealanders, Greeks, Germans, British, South Africans, and Turks [4]. There is currently a lack of clarity regarding whether the behavior patterns of Sri Lankan millennials align with those of other consumers in different countries, or if they exhibit distinct characteristics when making market decisions. As a result, this study aims to address this gap by conducting an original investigation into the decision-making styles of Sri Lankan millennials, which will be a pioneering effort in this area.

While previous research has provided extensive knowledge on consumer decision-making styles (CDMS), a paucity of studies focus solely on the impact of various factors on these styles. For example, gender [6], domestic brand biasness [3], students' background and family [7], Religiosity Commitment [8]. Hence, it is essential to identify the environmental conditions and factors that may influence the decision-making styles of millennials, a particular generation of consumers. Among the current approaches to influence millennial decision-making patterns, practitioners and scholars have placed particular emphasis on nostalgic marketing and neophilic tendencies.

The millennial features described above, such as their high degree of technical expertise, socially linked nature, and so on, appear to indicate a range of potentially fascinating collateral relationships to be examined with neophilia. On the other hand, the concept of neophilia is mostly researched on the aspects such food [9, 10], animal [11] innovation [12], and psychology [13]. However, marketing studies related to neophilia were relatively scarce. Similarly, other millennial traits, such as the fact that they are not fundamentally digital natives, that their buying environment is a mix of digital and analog marketplaces, and so on, tend to suggest a potentially fascinating collateral link with nostalgia. On the other hand, the concept of nostalgia is mostly researched on the aspects such as branding, advertising, and consumer attitude. In addition, though research conducted on nostalgia marketing has steadily increased, some gaps in literature can still be found [14].

As a result, this study aims to prioritize millennial decision-making styles and establish a more rigorous understanding of the association of newly emerging factors, such as nostalgia and neophilia, that have not before been examined alongside consumer decision-making styles. The researchers devised the following study objectives to attain the aforementioned research goal:

RO1: To investigate the types of decision-making styles of millennial consumers exercised in Sri Lanka.

RO2: To prioritize the types of decision-making styles of millennial consumers exercised in Sri Lanka.

RO2: To measure the association of Nostalgia with decision-making styles of Sri Lankan millennial.

RO3: To measure the association of Neophilia with decision-making styles of Sri Lankan millennial.

The introduction section of this chapter has established a foundation for the rest of the study by outlining the research problem, research objectives, and the importance of the research. The subsequent sections of this chapter are structured as follows. Firstly, a concise review of the existing literature pertaining to the research objective is presented, emphasizing key theories and empirical findings. Secondly, the research design and methods utilized in this study are described in detail, encompassing data collection, analysis, and interpretation. Finally, the limitations of the study and possible future directions are discussed.

2. Literature review

This section is a presentation of a review of literature pertaining to the key terms involved in the study: consumer decision-making styles (CDMS), millennial, nostalgia, and neophilia. The section begins with a discussion about consumer decision-making styles. The next part provides a comprehensive review of the literature on the constructs of Nostalgia and Neophilia. Further discussion includes the ways in which these two concepts connect with the millennial decision-making styles. Finally, the hypothesis of the study is presented.

2.1 Consumer decision-making styles

2.1.1 Overview of consumer decision-making styles (CDMS)

A consumer decision-making style is defined as "a mental orientation characterizing a consumer's approach to making choices" [15]. There are three (3) approaches to characterizing consumer decision-making styles in the consumer literature; the psychographic approach, the consumer typology approach, and the consumer characteristics approach.

According to psychographic research, consumer activity, personal interest, and viewpoint may be highly beneficial in gauging customer personalities and anticipating consumer behavior since they are more closely related to the purpose of marketing. The approach to consumer typology attempts to define "types" of the general consumer. However, this approach was criticized because they were skeptical that customers could be classified into separate unidimensional types of behavior [16]. Furthermore, he contended that this approach does not reflect the behavior of the market's growing "hybrid customer." The consumer characteristics approach is concerned with a variety of cognitive and emotive parameters/dimensions of consumer decision-making [15]. Consumer researchers have largely recognized the customer characteristics approach as the most interpretive and powerful construct among these approaches since it focuses on a cognitive and effective component of consumer behavior [17]. Sproles developed a tool for profiling consumer decision-making styles in the framework of the consumer characteristics approach in 1985. In 1986, Sproles and Kendall refined the existing inventory and developed the Consumer Style Inventory, a more logically consistent scale (CSI). The development of this refined inventory followed four key standards. Firstly, it includes mental characteristics that inform a consumer's decision-making process. Secondly, the inventory aims to be as comprehensive as possible, identifying only a limited number of basic and independent consumer decision-making traits. Thirdly, the inventory measures how consumers rate on each of these characteristics. Finally, the inventory takes into account measures that are of significance to consumer interest professionals [15].

2.1.2 Consumer style inventory (CSI)

The CSI included 40 items on consumer decision-making and measured the most basic eight [8] mental characteristics of consumer decision [15].

- Perfectionist/High-quality conscious consumer: This sort of customer seeks the greatest level of product quality. It is also possible that those consumers who are more quality sensitive may purchase more carefully, analytically, and methodically or by comparison.

- Brand-conscious consumer: These customers buy far more expensive well-known premium brands. Consumers who are more brand conscious assume that higher quality equates to a higher price.

- Novelty-fashion consciousness: Consumers who scored high on this aspect are more likely to become enthusiastic and prefer searching for new and stylish items. They keep up with trends, and it is essential for them to be fashionable. It also indicates that seeking variety is a key element of this trait.

- Recreational/hedonistic shopping consciousness: Consumers that belong to this category enjoy shopping. Previous research, however, has labeled this orientation as a "shopping avoider," thereby negatively charging most items.

- Price-conscious/"value for money" shopping consciousness: These customers are seeking for selling prices and are often interested in cheaper, discounted, or sale prices. Most importantly, customers get the most value for their money.

- Impulsiveness/Careless consumer orientation: These customers are unconcerned with shopping and never plan their purchases. They also appear unconcerned about their spending habits or the "best buys."

- Confused by over choice consumer: These customers perceive a plethora of brands and stores to select from, making it difficult to reach a choice. Furthermore, as several items in this factor indicate, they suffer from information overload.

- Habitual/Brand-loyal consumer: These customers are likely to have favorite brands and buying habits. The most well-recognized aspect of consumer decision-making is habitual behavior, which enhances its existence as a general trait.

The Consumer Shopping Inventory (CSI) has been tested and verified in various nations in order to get a better knowledge of consumer decision-making styles across cultures. Many experts, however, have challenged the instrument's applicability and generalizability for focusing primarily on a population of students that does not reflect the entire consumer continuum from diverse demographics and cultures [7, 17–19]. However, some of the results of the research are mainly consistent with the findings of the first CSI research [18]. However, a number of studies published in peer-reviewed journals show that substantial variations do exist; however, only few of them entirely reproduced all eight dimensions as exactly as the original CSI. As a result, the eight consumer decision-making styles have limited cross-cultural generalizability. Furthermore, given the numerous variances across the nations analyzed, the same traits may not be observed in Sri Lanka. As a result, the researcher has produced the following research.

RQ1: what are the types of decision-making styles of millennial consumers exercised in Sri Lanka?

RQ2: What are the most significant decision-making styles of millennial consumers exercised in Sri Lanka?

2.2 Nostalgia and Neophilia

2.2.1 Types of nostalgia

Term Nostalgia is a Greek word. Nostos means "return home," while algia refers to a painful ailment. As a result, nosatalgia is characterized as "a painful longing to return home" [19]. Over the last few decades, the concept of nostalgia has gained traction in consumer research [20]. According to the definitions of nostalgia, it reveals that slight changes can be visible in the definitions. But basically they are dealing with pleasant feelings toward the past or longing for something from the past.

It is possible to identify four types of nostalgia through literature, namely: Personal Nostalgia, Interpersonal Nostalgia, Cultural Nostalgia, and Virtual Nostalgia [21]. Personal nostalgia is real nostalgia based on personal experience that is triggered by genuine stimuli. Interpersonal Nostalgia is based on personal nondirect experience and memory [21]. This type of nostalgia is related to a time before birth and is transmitted by our parents or relatives through stories [22]. Cultural Nostalgia, also named as collective nostalgia, is based on the symbols of a particular. It simply refers to a group memory of Virtual Nostalgia/Historic Nostalgia based on books, video materials, and other nondirect experiences of a group of memory [21]. The researchers of this study focused on millennials' experience of nostalgic feelings toward their own past rather than the cultural and historical nostalgia.

2.2.2 Millennial and nostalgia

Havlena and Holak observed in 1991 that some generations are more prone to trigger nostalgic memories than others. He also said that each generation has its own set of symbols associated with nostalgic memories. Millennial consumers are the most responsive to nostalgic stimuli among the four main target groups in nostalgic marketing, namely experienced old consumers, special experience consumer groups, and consumer groups away from previous environments [21]. Similarly, millennials respond strongly to nostalgia as a result of their birth into a world of fast-changing technology. Furthermore, due to their present life stage of transition, millennials are psychologically at an opportune spot for this reflecting back [23]. Therefore, in this research study focuses on the millennial generation who faced to rapid social changes and the enormous psychological pressure which made them feel confused and uneasy [21]. Further, nostalgic components appear to be potentially useful tools for millennial consumers [24].

2.2.3 Neophilia

Neophilia refers to an inclination or attraction toward new or innovative things or experiences, as per its etymology, where "neo-" means new and "-philia" means attraction [23]. Merriam-Webster dictionary (2010) describes neophilia as a love or enthusiasm for new or novel things. Individuals who exhibit high levels of neophilia tend to be more receptive to novel ideas and products [25]. Louis Janda (2001) characterizes neophilia as an appreciation for and a desire to seek fresh and new experiences or a fondness for what is new. In essence, neophilia represents a preference for novelty and innovation. However, from a marketing standpoint, Neus Soler, a Professor at the University of Oberta de Catalunya, offers a comprehensive analysis of the rising trend of neophilia. She argues that neophilic customers are often leaders in their field, influential in the adoption of new technology and innovations, and indicators of the success of new products or services. Additionally, Nielsen Consulting underscores the importance of identifying neophilia as a critical factor in the survival of a new product.

2.2.4 Neophilia and millennial

There are many unique characteristics shared by the Millennial, such as novelty-seeking, receptive to innovation, technology-addicted (savvy), and trend follower [26]. Further, Millennial are the first to test out the new computer software, invest in the newest smartphone or open their minds to the world of artificial intelligence. Millennial is the generation that easily embraces the postworld on the Internet [27]. It indicates that millennial is a generation that accepts the novel and embraces changes. Due to the above-mentioned millennial characteristics, including high level of technological knowledge, their socially connected nature, and so on, seem to highlight a set of potentially interesting collateral relationships to be investigated alongside neophilia.

2.3 Nostalgia, Neophilia, and decision-making styles of millennials

According to reference (24), there is a strong correlation between nostalgia and consumer decision-making styles. The study found that individuals with a high level of nostalgia proneness had a significant positive association with perfectionism shopping, fashion-conscious shopping, and habitual shopping, as indicated by significant Pearson coefficients. Other previous research studies have also demonstrated that nostalgia can impact consumer behavior in various ways. For instance, one study conducted in 2012 found that ad-evoked nostalgia is linked to consumers' responses [28], while another study revealed that nostalgic cues in advertising could affect attitudes toward the advertised brand and advertising itself [28]. Moreover, previous research indicates that nostalgic feelings may influence consumption behavior. Furthermore, research has found that nostalgia has a significant positive effect on the perceived value, satisfaction, and loyalty to the destination [29]. Given these findings, it is evident that there could be several compelling secondary relationships between consumer decision-making styles and nostalgia [30]. Similarly, numerous studies have shown that neophilia can also influence consumer behavior [13, 32].

Based on the existing literature, the following research questions and hypothesis are proposed:

RQ3: Does nostalgia has a positive influence on one or more dimensions of decision-making styles of Millennial?

H1: Nostalgia has a positive influence on one or more dimensions of decision-making styles of Millennial.

RQ4: Does neophilia has a positive influence on one or more dimensions of decision-making styles of Millennial.

H2: Neophilia has a positive influence on one or more dimensions of decision-making styles of Millennial.

3. Research methodology

This section explicates the methodology which was used by the researcher in conducting the current study. According to the "Research Onion Concept", a researcher should design his research methodology like peeling off an onion from outer layer to inner layers [31]. By taking this concept into consideration, this section starts with the research philosophy then is followed by the research approach, research strategy, time horizons, and data collection methods, and finally, data analysis methods.

3.1 Research design

The current study incorporates the use of existing theories, such as consumer decision typologies, to construct hypotheses that will be evaluated during the research process, as is common in positivist philosophy. The survey was chosen as the research strategy, and the mono technique was chosen as the research method for the study.

The population for this study is defined as millennial consumers living in Sri Lanka. The convenience sampling technique was adopted, and a sample size of 385 was determined using Krejcie and Morgan's (1970) sample determination table, where the present study population is around 6.6 million (Mid-year Population Estimates by Department of Census and Statistics Sri Lanka., 2018). To obtain data on CDMS, nostalgia and neophilia self-administered questionnaires were devised. The questionnaire consisted of four sections, with the first section measuring consumer decision-making styles through 40 questions adapted from the original CSI scale by Sproles and Kendall (1986). The second section obtained data on nostalgia through eight questions adapted from the measurement scale of Perusson (2003) refined by Hajlaoui (2014). The third section measured neophilia with 12 questions taken from Walker & Gibbins' original neophilia scale (1989). The final section collected demographic information. A pilot test was conducted to ensure the face validity of the research instrument by giving the draft questionnaire to a few target respondents and experts.

In addition, another structured Questionnaire was created for the pairwise comparison format, with a scale of 1–9 being used, as shown in **Table 1**. Five experts

Intensity of Importance	Definition
1	Equally important
3	Moderately important
5	Strongly important
7	Very strongly important
9	Extremely strongly important
2,4,6,8	Intermediate values

Table 1.
AHP comparison scale.

were chosen for this part to provide pairwise comparisons for the AHP analysis. The experts included:

- A marketing researcher with a PhD in marketing and over 10 years of experience in consumer research and analysis, particularly among millennial consumers.
- A retail industry consultant with extensive experience analyzing millennial consumer behavior and preferences.
- An academic with a PhD in marketing and expertise in consumer behavior research.
- A consumer behavior expert with a master's degree in marketing and over 5 years of experience conducting consumer research and analysis.
- A millennial consumer representative who can provide firsthand insight into the decision-making process of the millennial generation.

3.2 Data analysis methods

Factor Analysis was used to determine whether the CDMS recognized by prior researchers were common to the sample of the Sri Lankan millennial. Further, it was used to reduce and summarize data and to identify underlying dimensions.

An Analytical Hierarchy Process (AHP) Approach was used to identify the most significant decision-making styles of millennial consumers exercised in Sri Lanka. The AHP is a simple decision-making technique developed by Saaty in 1980 for dealing with complicated, unstructured, and multi-attribute problems (Albayrak & Erensal, 2004).AHP is an effective decision-making method especially when subjectivity exists and it is very suitable to solve problems where the decision criteria can be organized in a hierarchical way into subcriteria [33].

The AHP consists of three main steps: The first step of the AHP consists of developing a hierarchical structure of the assessment problem. The overall goal is the solitary factor at the top of the hierarchy. The elements that influence the decision are referred to as attributes or criteria. Alternatives are decision options at the bottom of the hierarchy [32]. The author of the study conducted an empirical investigation and developed a hierarchical model of decision-making styles among millennials. The overall goal is to prioritize the consumer decision-making styles to identify the most significant element or attribute that influences consumer decision-making. Through an extensive review of literature, the study identified seven decision-making styles as criteria and evaluated each of them based on corresponding indicators, which were subcriteria or alternatives (**Table 2**) [15, 33].

In the second phase, once the problem has been deconstructed and the hierarchy has been established, the prioritization technique begins in order to identify the relative importance of the elements within each level. The prioritization method works by assigning a number from a comparison scale (**Table 1**) developed by Wind and Saaty (1980) to indicate the relative importance of the criteria.

The pairwise judgment begins at the second level (the first level of criteria) and ends at the lowest level, alternatives. As stated in the research design section for this study, pairwise comparison judgments were made by selected experts in the field using a structured questionnaire. The mathematical process for normalizing and determining the priority weights for each matrix begins in the third phase.

The consistency of the pairwise comparison assessments is directly related to the quality of the AHP output. Hence, it is critical to evaluate the levels of consistency in the comparison matrices [34]. Accordingly, the result of the comparison matrices was validated through consistency testing to avoid the subjective judgment that would lead to inaccurate results, as suggested by Saaty, 1982.

Eq. (1) consistency index expresses deviations from consistency, and the consistency index is a measure of inconsistency (CI) [32].

$$CI = \frac{\lambda_{\max} - \mathrm{n}}{n-1} \tag{1}$$

Eq. (2) is used to calculate the consistency ratio (CR), which is used to calculate a direct measure of pairwise comparisons' consistency [32]. The CR is calculated as the ratio of the CI and the random index (RI), (**Table 3**) where the values of the RI based on the number of evaluated criteria, as reflected in **Table 4**.

$$CR = \frac{CI}{RI} \tag{2}$$

The accepted upper limit for CR is the value 0.1. If the final CR is more than this value, the evaluation procedure must be repeated to ensure consistency [33]. To examine the third and fourth study questions, which were to identify the associations between nostalgia and neophilia with each CDMS dimension Pearson's Correlation Analysis was used.

4. Data presentation and analysis

4.1 The data preparation and sample profile

The data preparation process, which includes checking, editing, coding, transformation, data cleaning, systematic data adjustment, and finally, the selection of a

Factor	Code	Item	Loading	Eigenvalue	%Variance Explained	Cronbach's Alpha
Perfectionist Conscious	PC1	Getting very good quality is very important to me.	.796	6.623	22.076	.824
	PC2	When it comes to purchasing products, I try to get the very best or perfect choice	.816			
	PC3	In general, I usually try to buy the best overall quality	.671			
	PC4	My standards and expectations for products I buy are very high.	.676			
	PC5	I really give my purchases much thought and care	.715			
	PC6	A product should be perfect, or the best, to satisfy me.	.526			
Brand Conscious	BF2	The more expensive brands are usually my choice	.491	2.925	9.752	.828
	BF3	The most advertised brands are usually very good choices.	.813			
	BF4	The higher the price of a product, the better its quality.	.789			
	BF5	Nice department and specialty stores offer me the best products.	.765			
	BF6	I prefer buying the best-selling brands	.663			

Factor	Code	Item	Loading	Eigenvalue	%Variance Explained	Cronbach's Alpha
Novelty-Fashion Conscious	NC1	I keep my wardrobes up to date with the changing fashions.	.689	2.201	7.335	.778
	NC2	I usually have one or more outfits of the very newest style.	.831			
	NC3	Fashionable attractive styling is very important to me.	.734			
	NC4	To get variety, I shop at different stores and choose different brands.	.638			
Hedonistic & Recreational Conscious	HR1	Shopping is a pleasant activity for me	.874	1.966	6.554	.893
	HR2	Going shopping is one of the most enjoyable activities of my life.	.889			
	HR3	I enjoy shopping just for the fun of it	.814			
Habitual & Brand-Loyal Conscious	HB1	I have favorite brands I buy over and over	.690	1.502	5.005	.725
	HB2	Once I find a product or brand I like, I stick to it	.752			
	HB3	I go to the same store each time I shop	.697			
	BF1	The well-known brands are best for me.	.537			

Factor	Code	Item	Loading	Eigenvalue	%Variance Explained	Cronbach's Alpha
Impulsive, Careless Consumer	IC1	I should plan my shopping more carefully than I do.	.634	1.476	4.920	.658
	IC2	I am impulsive when purchasing.	.754			
	IC3	Often I make careless purchases I later wish I had not.	.695			
	IC5	I do not carefully watch how much I spend.	.659			
Confused by over choice	CO2	All the information I get on different products confuses me.	.760	1.290	4.300	.675
	CO3	The more I learn about products, the harder it seems to choose the best.	.776			
	CO4	Sometimes it is hard to choose which stores to shop.	.722			

Table 2.
Rotated component matrix for CSI.

data analysis strategy [35] was adopted by the researcher for the present study. Once the data has been prepared for the analyst, analyze the respondent's profile in order to understand and explain the characteristics of the particular sample. In the present study, the researcher used gender, educational level, and average monthly income level as demographic variables to analyze the sample characteristics shown in **Table 5**.

4.2 Factor analysis and goodness of measurement

The best method for factor analysis was then selected. The current study used principal component analysis, which took into account the entire variation in the data. Following the procedure of factor analysis, determine the number of factors. Among the various approaches, the study adopted based on eigenvalues combined with the percentage of variance.

The eigenvalues represent the amount of variation associated with the factor. As a result, only factors with eigenvalue variances greater than 1.0 are retained [35].

n	1	2	3	4	5	6	7	8	9	10
RI	0	0	0.58	0.9	1.12	1.24	1.32	1.41	1.45	1.49

Table 3.
Random index.

Variable	Dimensions	Tests with Standards			
		KMO	BTS Sig	CR	AVE
		0.5<	0.05>	0.7<	0.5<
Consumer Decision-Making Styles	Perfectionist Conscious	.841	.000	0.873	.536
	Habitual & Brand-Loyal Conscious	.725	.000	0.841	.570
	Novelty-Fashion Conscious	.750	.000	0.858	.603
	Brand Conscious	.828	.000	0.867	.568
	Hedonistic & Recreational Conscious	.726	.000	0.934	.825
	Confused by over choice	.656	.000	0.822	.606
	Impulsive, Careless Consumer	.686	.000	0.799	.498
Nostalgia	Reminder of Memories	.769	.000	0.876	.641
	The Past Regret	.784	.000	0.868	.622
Neophilia	Acceptance of Newness	.773	.000	0.812	.469
	Openness to Newness	.764	.000	0.839	.512

Table 4.
Goodness of measurements.

Sample description		f (N = 385)	%
Gender	Female	222	58%
	Male	163	42%
Income distribution	25,000 and below	253	66%
	25,000-50,000	61	16%
	50,000-75,000	20	5%
	75,000-100,000	17	4%
	100,000-125,000	12	3%
	Above 125,000	22	6%
Educational level	Ordinary level	2	1%
	Advanced level	5	1%
	Undergraduate	274	71%
	Degree	52	14%
	Postgraduate	52	14%

Table 5.
Sample profile.

The percentage of variation retrieved by the factors, on the other hand, should be sufficient. It is suggested that at least 60% of the variance be retrieved by factor [35]. In this work, the Varimax approach is employed as a rotation method to decrease the number of variables on a factor with high loads. **Table 2** illustrates the varimax rotated factor loading of the 40-item CSI, whereas **Tables 6** and **7** show the nostalgia and neophilia rotated component matrix for the Sri Lankan millennial sample, respectively.

Factor	Item Code	Item	Loading	Eigenvalue	%Variance Explained	Cronbach's Alpha
Reminder of Memories	NO1	In general, I remember my memories.	.857	3.307	41.343	.802
	NO2	Memories of the past come to my mind.	.895			
	NO3	I saw in my thought some of my memories.	.794			
	NO4	I think back to times of my life that are gone by now.	.577			
Past Regret	NO5	I regret not being able to relive my happy memories.	.781	1.823	22.782	.797
	NO6	I regret not being able to return to my past.	.824			
	NO7	I look back with regret that the right times of the past will not recur.	.807			
	NO8	I wish I could relive the emotions of the past.	.689			

Table 6.
Rotated component matrix for nostalgia.

Factor	Item Code	Item	Loading	Eigenvalue	% Variance Explained	Cronbach's Alpha
Openness to Newness	NE3	I think Pigs can fly	.776	2.601	26.014	.761
	NE4	I would like to be one of the first passengers to go to the moon	.716			
	NE5	More people ought to experiment with "mind-altering" drugs	.679			
	NE6	I would like to try eating a insects	.762			
	NE7	Crime is caused by the situations people find themselves in	.623			
Acceptance of Newness	NE1	I think there should be more changes in our society	.501	2.421	24.215	.709
	NE9	I like change	.732			
	NE10	Living in another country is a worthwhile experience	.648			
	NE11	People should always seek personal growth	.736			
	NE12	I make things in a way I like to use	.751			

Table 7.
Rotated component matrix for Neophilia.

Further, it is vital to ensure that the instrument used to measure a particular concept measures the variables accurately and that, in fact, the concept set out to measure is actually measured [36]. The goodness of the measures used in the study was therefore assessed by testing the reliability and validity of the instruments. Thus, all instruments comply with their tests and standards (**Table 4**). It has therefore been established that the instrument is valid and reliable. In addition, it is important that the data meet the basic assumptions underlying the statistical techniques to be used before proceeding with the data analysis [37] because analysis involves the use of a sample, not the population, and these assumptions form the basis of all multivariate statistical techniques. Since the requirement for normality, linearity, homoscedasticity, and multicollinearity has been met by all variables, it can be concluded that the data is suitable for multivariate analysis.

The rotated component matrix (**Table 2**) indicates that 29 items were successfully loaded into seven factors, each having an eigenvalue greater than 1.0. These factors accounted for approximately 60% of the total variance. As per the results presented in **Table 2**, the factor "price-value conscious" was removed, leaving seven factors. The researchers opted to name these remaining factors in a manner consistent with the original study conducted by Sproles and Kendall (1986) to maintain the similarity of the decision-making characteristics represented by each factor.

According to the findings of this study, the Sri Lankan millennial has seven decision-making styles: Perfectionist Consciousness, Habitual & Brand-Loyal Consciousness, Novelty-Fashion Consciousness, Brand Consciousness, Hedonistic & Recreational Consciousness, Confused by over choice, and the Impulsive Consumer.

4.3 Analytical hierarchical process

Then, to identify the most significant decision-making styles of millennial consumers exercised in Sri Lanka influence of the extracted factors of young consumer, the mathematical tool analytical hierarchical process (AHP) is used. The author has developed the decision hierarchy design first, followed by the prioritization process using pairwise comparison scale. **Table 8** shows the overall priority scores of millennial decision-making styles with factors priority within group with their percentages, rank within factors, and results of factors consistency tests.

CDMS		Priority	Rank	Results of consistency tests	Overall rank
			Perfectionist Conscious		
1	PC1	34.60%	1	Number of comparisons = 15	1
2	PC2	20.90%	2	Consistency Ratio CR = 0.6% (0.006 < 0.1)	
3	PC3	19.90%	3	Principal eigenvalue = 6.039	
4	PC4	6.80%	5	CI = 0.007, RI =1.24	
5	PC5	11.50%	4		
6	PC6	6.20%	6		
			Brand Conscious		
1	BF2	9.50%	4	Number of comparisons = 10	2
2	BF3	8.80%	5	Consistency Ratio CR = 1.5% (0.015 < 0.1)	
3	BF4	15.50%	3	Principal eigenvalue = 5.065	
4	BF5	26.10%	2	CI = 0.016, RI =1.12	
5	BF6	40.00%	1		
			Novelty-Fashion Conscious		
1	NC1	13.80%	3	Number of comparisons = 6	4
2	NC2	12.80%	4	Consistency Ratio CR = 0.4% (0.004 < 0.1)	
3	NC3	25.60%	2	Principal eigenvalue = 4.010	
4	NC4	47.80%	1	CI = 0.003, RI =0.9	

CDMS		Priority	Rank	Results of consistency tests	Overall rank
			Hedonistic & Recreational Conscious		
1	HR1	52.80%	1	Number of comparisons = 3	6
2	HR2	33.30%	2	Consistency Ratio CR = 5.6% (0.056 < 0.1)	
3	HR3	14.00%	3	Principal eigenvalue = 3.054	
				CI = 0.032, RI =0.58	
			Habitual & Brand-Loyal Conscious		
1	HB1	42.40%	1	Number of comparisons = 6	4
2	HB2	22.70%	2	Consistency Ratio CR = 0.4% (0.004 < 0.1)	
3	HB3	12.20%	4	Principal eigenvalue = 4.010	
4	BF1	22.70%	2	CI = 0.003, RI =0.9	
			Impulsive, Careless Consumer		
1	IC1	46.50%	1	Number of comparisons = 6	3
2	IC2	17.90%	3	Consistency Ratio CR = 3.6% (0.036 < 0.1)	
3	IC3	29.30%	2	Principal eigenvalue = 4.098	
4	IC5	6.30%	4	CI = 0.032, RI =0.9	
			Confused by over choice		
1	CO2	20.00%	3	Number of comparisons = 3	7
2	CO3	40.00%	1	Consistency Ratio CR = 0.0% (0.00 < 0.1)	
3	CO4	40.00%	1	Principal eigenvalue = 3.000	
				CI = 0.00, RI =0.58	

Table 8.
Overall consumer decision-making styles with AHP.

The study aimed to prioritize consumer decision-making styles using a hierarchical model depicted in **Figure 1**. The judgment matrix was used to compare the importance of each item in the data, and EXCEL was used to derive solutions through the weighted average method. To ensure consistency, a consistency test was conducted using Eq. (2), and all CR values were found to be lower than 0.1, indicating that all judgments were consistent. The weights of the factors were calculated based on the judgment matrix, and the total influence factors at the C level were evaluated. A weighted average was then used to rank each statement within their respective factor, as shown in **Figure 1**.

The results revealed that Sri Lankan millennials frequently followed three main decision-making styles – perfectionist consciousness, brand consciousness, and impulsive, careless decision-making, which explained 36.75% of the variance out of 60% variance in factor analysis.

4.4 Nostalgia and Neophilia with CDMS dimensions

Pearson's correlation coefficient was utilized to examine the association between neophilia and nostalgia with each dimension of consumer decision-making styles. **Table 9** shows that all seven CSI model dimensions appear to be positively correlated

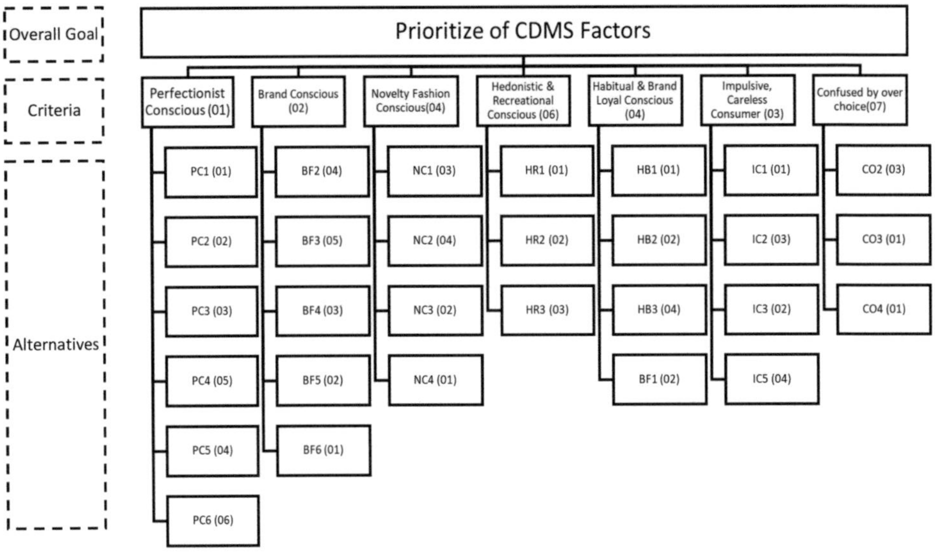

Figure 1.
Hierarchical structure of CDMS.

with nostalgia and neophilia, which has led to accept hypotheses H1 and H2. Perfectionist Conscious, Novelty-Fashion Conscious, Hedonistic & Recreational Conscious, and Impulsive, Careless Consumer decision-making styles are highly correlated with Neophilia while Habitual & Brand-Loyal Conscious, Brand Conscious, and Confused by over choice decision-making styles are highly correlated with nostalgia.

5. Discussion and conclusion

This chapter begins with a discussion of the key findings in respect to the research questions. Following this, general conclusions are made based on the study's findings. Finally, the implications, limitations, and recommendations for further research are addressed at the end of the section.

5.1 Key findings

The findings of the study revealed that Sri Lankan millennials exhibit seven distinct decision-making styles, including Perfectionist Consciousness, Brand Consciousness, Novelty-Fashion Consciousness, Hedonistic and Recreational Consciousness, Impulsive/Careless Consumer, Confusion from Overchoice Consumer, and Habitual and Brand-Loyal Consciousness. The research has further revealed a significant absence of price-value consciousness in the decision-making style of Sri Lankan millennials. The findings, based on the Analytic Hierarchy Process (AHP) technique, indicate that the perfectionist conscious decision-making style received the highest ranking in this study.

The fourth and fifth research objectives of this study were designed to investigate how nostalgia and neophilia influence the decision-making styles of Sri Lankan millennials. The findings suggest that both nostalgia and neophilia have a significant impact on the way these consumers make choices, highlighting the importance

		Perfectionist Conscious	Habitual & Brand-Loyal Conscious	Novelty-fashion conscious	Brand Conscious	Hedonistic & recreational conscious	Confused by over choice	Impulsive, Careless Consumer
Neophilia	Pearson Correlation	.151**	.104**	.134**	.151**	.177**	.211**	.267**
	Sig. (2-tailed)	.003	.001	.009	.003	.000	.000	.000
Nostalgia	Pearson Correlation	.130*	.167*	.133**	.231**	.157**	.255**	.239**
	Sig. (2-tailed)	.011	.000	.009	.000	.002	.000	.000

Table 9.
Nostalgia and Neophilia with CDMS dimensions.

of these psychological factors in shaping their decision-making styles. The study revealed that the correlation between neophilia and Novelty-Fashion Consciousness, Impulsive, and Careless Consumer decision-making styles is significantly stronger than their correlation with nostalgia. Conversely, the study found that the decision-making styles of Habitual & Brand-Loyal Conscious, Brand Consciousness, and being Confused by Over Choice are strongly associated with nostalgia. In other words, the results suggest that nostalgia plays a greater role in shaping these decision-making styles among Sri Lankan millennials than neophilia does.

5.2 Discussion

5.2.1 The Sri Lankan millennial generation does not exhibit a Price-conscious decision-making style

The current study's finding of the absence of price-value consciousness among Sri Lankan millennials is in line with the outcomes of prior research conducted in different countries such as New Zealand (43), China [17], Iran (44), and India [6]. These studies also observed the lack of a price-conscious decision-making style in their respective countries. The consistency of this finding across multiple cultural and geographic contexts implies a broader trend among Millennials and highlights the need for businesses to reconsider pricing strategy when targeting millennial. On the other hand, early adopters or innovators appear to be less price sensitive than later consumers [38]. According to prior research [39, 40], millennials are early adopters, which could be one of the reasons they are not price-conscious.

Furthermore, there could be several possible reasons why the Sri Lankan millennial generation does not exhibit a price-conscious decision-making style. One reason could be the cultural values and beliefs that prioritize quality and status over price. Sri Lankan society places a strong emphasis on social hierarchies, and individuals often use their purchasing decisions to signal their status and social position. Another reason could be the growing affluence of the Sri Lankan middle class, which has led to a shift in consumer preferences toward higher-end products and services. With more disposable income, millennials may be more willing to pay a premium for products that offer higher quality or better brand recognition [41]. Additionally, the rise of e-commerce and digital marketplaces in Sri Lanka may have created a more competitive pricing environment, making it easier for consumers to compare prices and find the best deals. As a result, millennials may feel less pressure to prioritize price in their decision-making process. Finally, the outcomes of the current study show that Sri Lankan millennials are perfectionists. These perfectionist customers may be willing to pay a higher price for superior quality [34]. It is crucial to acknowledge that the aforementioned rationales are merely potential explanations, and a more comprehensive investigation is necessary to fully comprehend the variables that shape the decision-making approach of the Sri Lankan millennial cohort.

However, several scholars claimed that the price-conscious construct had certain reliability issues [42]. As a result, the price-conscious construct required refinement [38].

5.2.2 Millennial consumers in Sri Lanka exhibit a high degree of perfectionist consciousness

According to the AHP technique, the perfectionist conscious decision-making style received the highest rank in this study. The finding is similar to [4] prior research, which found perfectionist consciousness as the most popular style among Indian, Korean, Chinese, and American consumers. This is also one of the consistent decision-making styles that have been shown in a number of earlier studies [1, 6, 7, 18]. Moreover, there may be several possible factors that contribute to the high degree of perfectionist consciousness exhibited by Sri Lankan millennials. Such as, the emphasis on education and achievement in Sri Lankan culture may lead to a high degree of perfectionism among Millennials, including in their decision-making style. This focus on excellence and high standards may also extend to their standard of living, as Millennials may strive to achieve a certain level of material success and financial stability. Additionally, the influence of social media and the Internet may create a perception of what a desirable standard of living looks like, further reinforcing the importance of perfectionism and achieving high standards in decision-making and in life in general. Further findings revealed this style was followed by the Brand Conscious decision-making method. This finding is consistent with a study conducted using AHP analysis in India, where brand consciousness is ranked first [33].

5.2.3 The impact of nostalgia and neophilia on consumer decision-making styles

The results of this study indicate that Nostalgia and Neophilia impact the consumer decision-making styles of Sri Lankan millennials. The available evidence suggests that nostalgia and neophilia may positively influence consumer decision-making styles, although the extent of their influence may vary depending on the individual and the specific context.

The present study's findings are consistent with prior research, which has demonstrated a positive association between the decision-making styles of Generation Y consumers and their tendency toward nostalgia [7].

According to research, the association between neophilia and consumer behavior styles such as Perfectionist Conscious, Novelty-Fashion Consciousness, Hedonistic and Recreational Conscious, and Impulsive and Careless Consumer is stronger than their association with nostalgia. Perfectionist Conscious consumers are likely to exhibit a strong correlation with neophilia as they seek out high-quality and innovative products and are willing to pay a premium for such products. Similarly, Novelty-Fashion Conscious consumers may have a strong correlation with neophilia because they possess a keen interest in fashion and are willing to take risks with their fashion choices by adopting new trends. The Hedonistic and Recreational Conscious consumer may also have a strong correlation with neophilia due to their pursuit of pleasure and excitement, making them more likely to try new products that offer novel experiences. Lastly, Impulsive and Careless Consumer decision-making styles may exhibit a strong correlation with neophilia due to their tendency to make spontaneous and novelty-seeking purchases without much consideration for the long-term consequences.

Nostalgia is characterized by a sentimental attachment to positive memories of the past, which can influence consumer behavior. Consumers who exhibit Habitual & Brand-Loyal Conscious, Brand Consciousness and are Confused by over choice decision-making styles tend to have a stronger association with nostalgia. Habitual & Brand-Loyal Conscious consumers demonstrate a high correlation with nostalgia due to their long-term relationship with a particular brand or product, leading to positive brand associations and a sense of nostalgia. Similarly, Brand Consciousness is highly correlated with nostalgia due to the emotional attachment and positive memories associated with the brand image and reputation. Consumers who feel overwhelmed or anxious due to too many choices may exhibit a Confused by over choice decision-making style, which is also highly correlated with nostalgia. These consumers may seek comfort in familiar products or brands, leading to nostalgic feelings toward products or brands from their past. Overall, these findings are consistent with the profile of individuals prone to nostalgia, who have a prominent desire for the past. Hence, this finding is consistent with the profile of an individual prone to nostalgia, who has a prominent desire for the past [39].

5.3 Implications

This study's findings significantly contribute to the existing literature in several ways. Firstly, it addresses the need for the original form of the Consumer Style Inventory (CSI) to be reassessed in different countries. Additionally, the study is one of the few attempts to investigate consumer decision-making styles in Sri Lanka [39–41]. Therefore, this study provides valuable insights into the decision-making styles of Sri Lankan millennials, which can be useful for marketers and businesses targeting this demographic. Additionally, the findings can help researchers better understand the consumer behavior of Sri Lankan millennial and contribute to the development of more effective marketing strategies. Moreover, with the use of the multi-criteria decision-making technique AHP, this study provides new insight into identifying the most common CDMS of millennials in Sri Lanka. Furthermore, using AHP technique adds a methodological contribution to the study.

Second, upon a thorough review of the literature, it becomes apparent that there is a gap in research regarding the concepts of nostalgia and neophilia and their impact on consumer decision-making. Despite previous studies on nostalgia, there has been a lack of attention given to its influence on consumer decision-making. Moreover, to the best of our knowledge, no research has been conducted on the effect of neophilia on consumer decision-making styles. Additionally, most existing studies have not investigated the impact of both nostalgia and neophilia on consumer decision-making styles simultaneously. Therefore, this study's contribution to the existing body of knowledge is to address these gaps by exploring the relationship between nostalgia, neophilia, and consumer decision-making styles. This study's pioneering role can help fill these research gaps and provide valuable insights for marketers and businesses seeking to understand consumer behavior and develop effective marketing strategies. Furthermore, no research was found to investigate the impact of nostalgia and neophilia on millennial decision-making styles with reference to the context of Sri Lanka. Through this study, all above-mentioned knowledge gaps were filled.

The practical contribution of this study is primarily relevant to marketing managers. The study's findings can assist marketers in understanding the decision-making styles of Sri Lankan millennials, providing them with valuable insights to better target this generation. The Consumer Style Inventory (CSI) used in this study is a reliable and valid tool for analyzing millennial consumers' behavior in Sri Lanka, providing a useful tool for marketers to create consumer profiles to guide marketing strategy. Moreover, marketers can use the findings of this research to segment, target, and position their products or services more effectively. Overall, this study's practical contribution can help marketers overcome challenges in targeting Sri Lankan millennials by providing them with a better understanding of their decision-making styles, resulting in more effective marketing strategies and better business outcomes.

Moreover, examining the association between nostalgia and millennial decision-making styles has practical contributions for companies. They can leverage this knowledge to develop products and marketing campaigns that evoke positive emotions and memories from the past, thereby enhancing consumer engagement and brand loyalty. Understanding the role of nostalgia in decision-making can also guide product development by identifying design elements that trigger nostalgic feelings in consumers, resulting in increased purchase intent. In addition, businesses can create consumer experiences that leverage nostalgia to generate positive emotions and memories, ultimately leading to higher consumer satisfaction and engagement. Overall, researching the association of nostalgia with millennial decision-making styles can have practical implications for product development, marketing, and consumer experiences, enabling companies to formulate effective strategies to engage and appeal to this crucial demographic group. In addition, as per the research findings, businesses and marketers should consider incorporating new and innovative products or services into their offerings to appeal to consumers who have a strong neophilic inclination. For consumers who exhibit perfectionist-conscious behavior, marketers should emphasize the uniqueness and novelty of their products, while also highlighting the quality and attention to detail in their design and production. For those with novelty-fashion consciousness, businesses can focus on promoting cutting-edge and trendy styles, using social media platforms to showcase new arrivals and limited edition collections. Hedonistic and recreational-conscious consumers may be drawn to products that promise new and exciting experiences, such as adventure tourism, extreme sports, or high-tech gadgets that offer sensory stimulation.

Lastly, impulsive and careless consumers may be particularly susceptible to impulse purchases triggered by a desire for novelty and excitement. As such, businesses should take care not to exploit these consumers by marketing products in a way that encourages reckless spending. Understanding the connection between neophilia and consumer behavior styles can help businesses tailor their marketing strategies to appeal to a wide range of consumer preferences and increase their chances of success in the competitive marketplace.

5.4 Limitations and further research

The research is subject to certain limitations that warrant consideration when interpreting the findings and implications. Firstly, the study was restricted to millennial consumers residing in Sri Lanka, so the generalizability of the results to other generations and populations may be limited. Future investigations may benefit from a more diverse sample, encompassing respondents from various generations, in order to draw more robust and generalized conclusions. Additionally, as this was a quantitative study, its ability to fully comprehend the intricacies and processes underlying the interaction between nostalgia, neophilia, and consumer decision-making styles may have been restricted. Given that nostalgia and neophilia are concepts highly intertwined with consumer psychology, qualitative research would be more appropriate in order to enhance our understanding of these constructs.

6. Conclusion

In conclusion, this study has shed light on the decision-making styles of Sri Lankan millennials and the role of nostalgia and neophilia in shaping their choices. The findings based on AHP, indicate that Sri Lankan millennials exhibit various decision-making styles, with perfectionist consciousness being the most prominent.

Interestingly, the study found a significant absence of price-value consciousness in their decision-making styles. Furthermore, the study highlights the importance of nostalgia and neophilia in influencing their choices. The findings suggest that while both factors have a significant impact on decision-making, nostalgia appears to be more closely associated with certain styles, such as Habitual & Brand-Loyal Consciousness and being Confused by Over Choice. In contrast, neophilia is more strongly linked to Novelty-Fashion Consciousness, Impulsive, and Careless Consumer decision-making styles. Overall, this study provides valuable insights into the decision-making patterns of Sri Lankan millennials and the role of psychological factors in shaping their choices.

Author details

Vidana Gamage Pavani Lakshika[1] and M.J.M. Fazeela Ahzan[2*]

1 Department of Marketing Management, University of Sri Jayewardenepura, Nugegoda, Sri Lanka

2 Department of Marketing, University of Colombo, Sri Lanka

*Address all correspondence to: fazeelaja@mkt.cmb.ac.lk

References

[1] Lakshika V, Ahzan M. Decision making styles of Sri Lankan millennial: Are they different? Sri Lanka Journal of Marketing. 2021;**7**(1):46

[2] Bae S, Pyun D, Lee S. Consumer decision-making styles for Singaporean college consumers: An exploratory study. The International Council for Health, Physical Education, Recreation, Sport and Dance Journal of Research. 2010;**5**(2):70-76

[3] Wanninayake BWMC. Consumer decision-making styles and local brand biasness: Exploration in the Czech Republic. The Journal of Competitiveness. 2014;**6**(1):3-17

[4] Mokhlis S. An investigation of consumer decision-making styles of young-adults in Malaysia. International Journal of Business and Management. 2009;**4**(4):140-148

[5] Bolton RN, Parasuraman A, Hoefnagels A. Understanding generation Y and their use of social media: A review and research agenda. Journal of Service Management. 2013;**24**(3):245-267

[6] Sharma MK, Aich S. Consumer shopping decision making styles at departmental stores: An exploratory study of gender differences. Indian Journal of Commerce & Management. 2012;**III**(2):51-57

[7] Azizah Haron S, Chinedu AH. Consumer background and decision making styles of Malaysian college students. International Journal of Business and Management. 2017;**13**(1):170

[8] Aliman NK, Ariffin ZZ, Hashim SM. Religiosity commitment and decision-making styles among generation Y Muslim consumers in Malaysia. International Journal of Academic Research in Business and Social Sciences. 2018;**8**(1):554-576

[9] van Trijp HCM, van Kleef E. Newness, value and new product performance. Trends in Food Science and Technology. 2008;**19**(11):562-573

[10] Giordano S, Clodoveo ML, De GB, Corbo F. Factors determining neophobia and neophilia with regard to new technologies applied to the food sector: A systematic review. The International Journal of Gastronomy and Food Science. 2018;**11**:1-19. DOI: 10.1016/j.ijgfs.2017.10.001

[11] Brown GR, Nemes C. The exploratory behaviour of rats in the hole-board apparatus: Is head-dipping a valid measure of neophilia? Behavioural Processes. 2008;**78**(3):442-448

[12] Scott-Smith T. Humanitarian neophilia: The 'innovation turn' and its implications. Third World Quaterly. 2016;**37**(12):2229-2251. DOI: 10.1080/01436597.2016.1176856

[13] Elena L. A study regarding the Neophilia or tendency towards novelty for two types of students. Science, Movement and Health. 2015;**15**(2):127-134

[14] Holbrook MB, Schindler RM. Nostalgic bonding: Exploring the role of nostalgia in the consumption experience. 1996;**3**:107-127

[15] Sprole GB, Kendall E. A methodology for profiling consumers' decision-making. Journal of Consumer Affairs. 1986;**2**(20):267-279

[16] Walsh G, Hennig-Thurau T, Wayne-Mitchell V, Wiedmann K-P. Consumers' decision-making style as a basis for market segmentation. Journal of Targeting, Measurement and Analysis for Marketing. 2001;**10**(2):117-131

[17] Siu NYM, Wang CCL, Chang LMK, Hui ASY. Adapting consumer style inventory to chinese consumers: A confirmatory factor analysis approach. Journal of International Consumer Marketing. 2001;**13**(2):29-47

[18] Hanzaee KH, Aghasibeig S. Generation Y female and male decision-making styles in Iran: Are they different? The International Review of Retail, Distribution and Consumer Research. 2008;**18**(5):521-537

[19] Davis F. Yearning for yesterday: A sociology of nostalgia. New York: Free Press. 1979;**9**(3):410-411

[20] Preseren S, Koklic MK. The role of nostalgia in consumer decision making in Slovenia. In: Management International Conference. 2017. pp. 169-170

[21] Cui R. A review of nostalgic marketing. Journal of Service Science and Management. 2015;**8**(February):125-131

[22] Kessous A, Roux E. Brands considered as "nostalgic": Consequences on attitudes and consumer-brand relationships. Rech Appl en Mark (English Ed). 2010;**25**(3):29-55

[23] Hungate HL. The Rearview Mirror and Windshield: Utilizing the Intersection of Nostalgia and Neophila to Target Today's Consumers. Fayetteville, AR, USA: University of Arkansas; 2018

[24] Betti M, Jahandad I. The Unique Nostalgic Shopper. Sweden: Jonkoping University; 2016

[25] Li M, Frieze IH, Horvat J. Reasons for leaving home: Comparing predictors of wanting to migrate and travel in Croatian undergraduates. Migration and Ethic Themes. 2012;**28**:7-27

[26] Moreno FM, Lafuente JG, Carreón FÁ, Moreno SM. The characterization of the millennials and their buying behavior. International Journal of Marketing Studies. 2017;**9**(5):135

[27] Newman D, Kramer S, Blanchard O. Marketing to Millenials [Internet]. 2016. Available from: https://futurumresearch.com/wp-content/uploads/woocommerce_uploads/2017/02/report_200117.pdf

[28] Muehling DD, Sprott DE, Sprott DE. The power of reflection: An empirical examination of nostalgia advertising effects an empirical examination of nostalgia advertising effects. Journal of Advertising. 2013;**33**(3):25-35

[29] Verma A, Rajendran G. The effect of historical nostalgia on tourists' destination loyalty intention: An empirical study of the world cultural heritage site–Mahabalipuram, India. Asia Pacific Journal of Tourism Research. 2017;**22**(9):977-990

[30] Khodayari S, Younesi J, Barnaj AF. Modeling of attitude towards drug addiction among university students in Tehran. etiadpajohi. 2016;**10**(37):22-32

[31] Saunders M, Lewis P, Thornhill A. Research Methods for Business Students. Fifth ed. England: Pearson Education Limited; 2009

[32] Albayrak E, Erensal YNC. Using analytic hierarchy process (AHP) to improve human performance: An application of multiple criteria decision

making problem. Journal of Intelligent Manufacturing. 2004;**15**:491-503

[33] Kumar A, Dash MK. Factor exploration and multi-criteria assessment method (AHP) of multi-generational consumer in electronic commerce. International Journal of Business Excellence. 2014;**7**(2):213-236

[34] Sneesl R, Jusoh YY, Jabar MA, Abdullah S, Bukar UA. Factors affecting the adoption of IoT-based smart campus: An investigation using analytical hierarchical process (AHP). Sustain. 2022;**14**(14):1-21

[35] Malhothra NK, Nunan D, Birks DF. Marketing research. In: The Marketing Research: An Applied Approach. fifth ed. London, United Kingdom: Pearson Education Limited; 2016. p. 976

[36] Sekaran U, Bouige R. Research Methods for Business: A Skill Building Approach. 7th ed. West Sussex, United Kingdom: Wiley India Pvt Ltd; 2018. p. 407

[37] Hair JF, Black WC, Anderson RE, Babin BJ. Multivariate Data Analysis (MVDA). Seventh. Pharmaceutical Quality by Design: A Practical Approach. Essex, England, United Kingdom: Pearson Education Limited; 2017. pp. 201-225

[38] Siu NYM, Wang CCL, Chang LMK, Hui ASY. Adapting consumer style inventory to Chinese consumers. Journal of International Consumer Marketing. 2001;**13**(2):29-47

[39] Sierra JJ, Mcquitty S. Attitudes and emotions as determinants of nostalgia purchases: An application of social identity theory. Journal of Marketing Theory and Practice. 2014;**15**(2):99-112

[40] Toledo AC, Lopes EL, De JUN. Effect of Nostalgia on Customer Loyalty to Brand Post-Merger/Acquisition. 2015;**2016**(2016):33-55

[41] Central Bank of Sri Lanka. CENTRAL BANK OF SRI LANKA ANNUAL REPORT 2019 [Internet]. 2019. Available from: https://www.cbsl.gov.lk/en/publications/economic-and-financial-reports/annual-reports/annual-report-2019

[42] Anić ID, Ciunova-Shuleska A, Rajh S, Rajh E, Bevanda A. Differences in consumer decision-making styles among selected south-east European countries. Economic Research-Ekonomska Istraživanja. 2016;**29**(1):665-681. DOI: 10.1080/1331677X.2016.1193949

Chapter 3

Implementing MCDA to Determine Eligibility for Pardon

Irit Talmor

Abstract

Countries celebrating festive national events sometimes declare a wide-scale pardon of prisoners, based on certain criteria. This paper presents an original and effective method for producing a list of prisoners eligible for early release by utilizing a multi-criteria decision analysis (MCDA) approach. The following criteria were used: gender, age, severity of crime, proportion of sentence already served, and behavior in the last three years. Criteria weights were calculated using an analytical hierarchy process (AHP), based on data provided by the authorities. The model produced clear and unbiased results, offering policymakers a practical and user-friendly decision-making tool. It was implemented on a list of 1500 candidates for early release through a wide-ranging amnesty, and several hundred of them achieved scores that made them eligible for a pardon.

Keywords: pardon, criteria ranking, AHP, decision support tool, clarity

1. Introduction

An amnesty is a pardon granted to convicted prisoners. In countries marking events such as Independence Day, an amnesty policy could grant pardons to specific groups of prisoners or reduce their sentences. This is typically done as a goodwill gesture aimed at promoting national unity. The authorities aim to determine the policy of eligibility for clemency in a way that benefits both the prisoners and the public [1, 2]. Typically, the amnesty is preceded by complex bureaucratic and decision-making processes. Two important parameters that must be determined during this phase are the total number of prisoners to be freed and the set of criteria for early release. These criteria may include the severity of the crime, the portion of the sentence a prisoner has already served, their behavior while in prison, and the likelihood that they will re-offend if released. Formerly, the authorities made such multi-dimensional sensitive eligibility comparisons without using any systematic, objective method. Thus, they faced public criticism and anger.

This article demonstrates the use of two-dimension multi-criteria decision analysis (MCDA) as a simple method to help authorities decide which prisoners to pardon.

The MCDA approach is a popular and user-friendly tool for choosing among alternatives. As such, this technique is widely used to make decisions in diverse fields, such as choosing between investment alternatives [3], locating electric vehicle charging stations [4], evaluating the performance of inland ports [5], allocating budgets for political campaigns [6], and even assessing steam boiler alternatives [7]. The approach has also garnered a great deal of attention in the academic literature, as seen in [8–10]. The criteria weights – that is, their relative importance – were calculated using the basic principle of the Analytic Hierarchy Process (AHP), the pairwise comparisons [11, 12].

The method was implemented in Excel and produced clear and straightforward results in a convenient format that is easy to understand and to use.

The rest of the article is organized as follows: the methodology is explained in the next section; then, the results and conclusions are presented.

2. Methodology

Ahead of Independence Day, the authorities in a certain country announced a wide-ranging amnesty for prisoners. The decision-makers defined the following criteria for pardon: gender, age, severity of crime, proportion of sentence already served, and behavior in the last three years. In addition, the authorities wanted to consider the prisoner's dominance in prison as well as their stature in the local underworld.

Although the criteria were well defined, the method of translating them into a set of unbiased decisions had to be determined in order to produce the list of pardoned prisoners. Thus, the authorities were looking for a simple but reliable method to assist in decision-making, and MCDA was suggested as an effective model. We applied it using three stages: First, we decided on the criteria; second, we applied an AHP, comparing each criteria-pair to calculate the relative criteria weights. Third, we used these weights to produce a list of prisoners eligible for pardon. In the following paragraphs, we detail these stages.

2.1 Stage 1: setting the criteria

The criteria were set in cooperation with the authorities. We asked them to list the factors that they thought should be considered in the pardon process and explain their preferences. This stage entailed a few rounds of discussions and enabled us to finalize the criteria, as follows (in alphabetical order):

Age - the authorities preferred to grant pardons to prisoners that were younger than 21 and older than 67. This is because youngsters are prone to making impulsive decisions that they later regret [13]. As a result, they were seen as more amenable to rehabilitation and a lesser risk to public safety. On the other hand, older prisoners were seen as a lesser threat to public safety due to their advanced age and declining health. They were also seen as less likely to re-offend because of their age and the fact that they have already spent a significant portion of their lives in prison.

Behavior - the authorities preferred to grant pardons to prisoners who demonstrated exemplary behavior while in prison. This is because good behavior was seen as a sign that prisoners accepted responsibility for their actions, are remorseful, and are making an effort to change their ways. As a result, their odds of reintegration into society were considered high. The contrary also holds true: the authorities preferred not to grant early release to prisoners who exhibited unruly behavior in prison. This was interpreted as a failure to accept responsibility for wrongdoing, and a sign that the prisoners are not remorseful and are making no effort to reform [14, 15]. Thus, releasing such prisoners would not promote reintegration into society while posing a heightened risk to public safety.

Crime severity - the authorities were willing to grant early release to non-violent prisoners who committed non-violent crimes (e.g., white collar offenses, theft, etc.). Officials were less inclined to pardon prisoners who committed severe crimes. Specifically, prisoners with "blood on their hands" were excluded from the amnesty campaign. The authorities believed that releasing these prisoners would likely cause a public uproar, and that such offenders are more dangerous to society.

Gender - the authorities showed greater willingness to release females compared to males. This is because the number of female prisoners was much lower, and they were believed to pose a lower risk to public safety due to their limited physical strength and generally lower levels of aggression. In this context, the authorities viewed non-binary prisoners the same way as females.

Health condition – authorities preferred to grant early release to prisoners with serious health problems. This is because their declining health and limited strength usually decreased the threat they posed to public safety [16, 17]. Considerations such as the costs of medical treatment in prison and the risk of sudden deterioration or death while in jail also carried substantial weight for this preference.

Portion of sentence served - the authorities viewed the portion of the sentence a prisoner has already served as a relevant factor when deciding whether to grant them a pardon. Officials pointed out that the larger this portion was, the less they were concerned about the potential risk to public safety. More importantly, public opinion was more favorable toward releasing such inmates.

Hierarchies among inmates are common in prison [2]. Thus, the authorities wanted to add one more criterion: the prisoner's dominance within the hierarchy. That is, whether the prisoner is a leader who enjoys an elevated status among the other prisoners and has loyalists inside and/or outside the prison who follow his orders. Early release of such prisoners may have a broader effect on society given their leading role and ability to initiate and direct criminal activity. The authorities were initially unsure which policy to apply to this group, so it was essential to distinguish this criterion from the others.

The complete list of criteria, values and preferences is given in **Table 1**.

Criterion	Values	Preferences	Comments
Age	The prisoners' age, ranging from youngest to oldest	The authorities preferred to pardon younger and older prisoners	The ages of "younger" and "older" prisoners may vary, but must be below 21 and above 67
Behavior	Good Average Bad	Good behavior increased the odds of early release	The assessment was based on the prison records of inmates
Severity of crime	Minor offenses (white-collar, theft) Intermediate (robbery, drugs) Severe (murder, aiding and abetting murder)	The authorities excluded prisoners with "blood on their hands"	Some inmates who aided and abetted serious crimes were eligible for pardon, depending on their specific background

Criterion	Values	Preferences	Comments
Gender	Female Other Male	The authorities preferred to pardon females/other prisoners	
Health condition	Healthy Minor health problems Major health problems	Severe health problems increased the prisoner's odds of getting a pardon	The health condition was determined by the prison's doctor
Portion of sentence served	Ranged from 10–90%	Prisoners who served most of their sentence were more likely to be pardoned	Prisoners who served a negligible portion of their sentences were not eligible for a pardon. Prisoners who nearly completed their sentences were automatically pardoned
Dominance	• -High • -Medium • -Little or none	The preferences were not unequivocal	

Table 1.
List of criteria, values, and preferences (in alphabetical order).

2.2 Stage 2: weighing the criteria

The criteria were weighed using AHP. According to this method, the decision-makers determine their preference for each pair of criteria using a numeric scale ranging from 1 to 9 (see **Table 2**).

It is common to use odd numbers to distinguish among measurement points. The use of even numbers is adopted only as a compromise between evaluators. The preferences are collected in a matrix, as shown in **Figure 1**. In this example, there are three criteria: C1, C2, and C3. The decision-maker extremely prefers C1 to C2, moderately prefers C1 to C3, and strongly prefers C3 to C2. The matrix is reciprocal because the lower triangle is the reverse of the upper triangle, and its diagonal is filled with "1", comparing each criterion to itself.

Degree of preference	Numeric value
Equal	1
Equal to moderate	2
Moderate	3
Moderate to strong	4
Strong	5
Strong to very strong	6
Very strong	7
Very strong to extreme	8
Extreme	9

Table 2.
Saaty's scale of preferences [9].

	C1	C2	C3
C1	1	9	3
C2	1/9	1	1/5
C3	1/3	5	1

Figure 1.
Preferences matrix.

After completing the matrix, a consistency check is performed to detect possible contradictions and fix them. Such inconsistencies may occur when the problem is vaguely defined or when the evaluator faces difficulties in maintaining consistency while comparing too many pairs. As the number of pairs to compare quadratically increases with the number of criteria, achieving consistency is indeed a challenge. An acceptable CR (consistency ratio) value should be less than 0.1.

In our case, we had six criteria (excluding "Dominance"); thus, 15 comparisons were required. To simplify the process for decision-makers, we interviewed three relevant officials and asked each to compare 10 pairs, so each pair got two independent preferences. This technique enabled us to simplify the process for time-constrained decision-makers while still recognizing gaps (or points of agreement) in their preferences. The decision matrix (based on adjusted comparison results) is presented in **Figure 2**. The resulting weights are shown in **Table 3**. The values were calculated using [18].

Unexpectedly, the Portion of Sentence and Behavior criteria were marginalized, whereas the Gender criterion was given significant weight. We believe that this stemmed from official priorities. The officials estimated that policymakers wanted the broadest pardon possible, particularly for women. Thus, considerations such as Behavior or Portion of Sentence played a minor role in the decision, whereas gender played a major role.

	Age	Behavior	Crime severity	Gender	Health conditions	Portion of sentence
Age	1	5	0.14	0.25	0.25	2
Behavior	0.2	1	0.14	0.17	0.25	0.33
Crime severity	7	7	1	2	3	4
Gender	4	6	0.5	1	2	8
Health conditions	4	4	0.33	0.5	1	6
Portion of sentence	0.5	3	0.25	0.13	0.17	1

Figure 2.
Decision matrix.

Criterion	Rank	Weight
Crime severity	1	37.8%
Gender	2	27.1%
Health Conditions	3	18.6%
Age	4	7.9%
Portion of sentence	5	5.4%
Behavior	6	3.2%

Table 3.
Resulting ranks and weights.

2.3 Stage 3: selecting prisoners to pardon

Ranking the eligibility of prisoners for early release was performed as follows: First, we defined the possible scores for each criterion based on the preferences of the authorities (see **Table 4**). Second, each pardon candidate was graded for each criterion. Third, a prisoner's overall score was obtained by calculating the score for each criterion based on its weight and summing up the scores. The prisoners were then listed in descending order based on their overall scores, thus enabling the authorities to consider this score in tandem with the dominance ranking to make the pardon decision.

For example, the maximum score (100) in the Age criterion was given to the youngest and oldest prisoners, who happened to be 17 and 75 years old, respectively. The minimum grade (0) was given to age 46, which was the average between those two edges. Each year below/above 46 increased the grade by 100/29 points.

To illustrate the process, assume a male prisoner named John. John is 55 years old. He was sentenced to 20 years for robbery 12 years ago. In the last three years, he has been well-behaved and prison authorities have a positive opinion of him. Although suffering from diabetes, he manages to balance it. He is not dominant in jail and prefers to remain outside the predominant groups.

Based on John's data, we can determine his scores: He is 55, so his score for the Age criterion is approximately 69 (= 20*100/29); his score for Behavior is 100; his crime severity score is 50; his Gender score is 0. Similarly, John's Health Condition score is 50, and his Portion of the Sentence score is 60 (=12/20*100).

John's overall score is calculated by multiplying his grades by the criteria weights, as follows:

$$69 \cdot 7.9\% + 100 \cdot 3.2\% + 50 \cdot 37.8\% + 0 \cdot 27.1\% + 50 \cdot 18.6\% + 60 \cdot 5.4\% = 40.1\% \quad (1)$$

Note that for identical criteria values except for Gender, the score would be 67.2%. This gap clearly shows the authorities' strong preference for early release of female prisoners.

Criterion	Values	Possible scores
Age	17 to 75	17 or 75: 100 17 < x < 75: 100 – min(75-x, x-17)*100/29
Behavior	Good, Average, Bad	100, 50, 0
Crime severity	Modest, Intermediate, Severe	100, 50, 0
Gender	Female, Binary, Male	100, 100, 0
Health conditions	Major problems, Minor problems, No problems	100, 50, 0
Portion of sentence	Ranged from 10–90%	Time already spent in prison/total punishment*100
Dominance	High Medium Little or none	The preferences were not unequivocal

Table 4.
Range of scores for the first six criteria.

3. Results

The approach and the tool developed for it highly met the authorities' expectations. Thus, they decided to examine them on a list of potential prisoners for pardon, which included 1500 names. The highest score was 76.1 and the lowest was 4.2. The results map is shown in **Figure 3** and **Table 5**. These results show two dominant ranges: in the 51–60 range and 21–30 range. Whereas the low range had less significance because such scores were not eligible for early release, the high range posed a dilemma for the authorities: including those prisoners in the amnesty program meant an early release of more than 850 prisoners, far above the original allocation. But not including them meant releasing only 60 prisoners – far below the initial allocation. To the author's best knowledge, the solution was to set the threshold at a value between 41 to 50 and to deny a pardon to dominant prisoners.

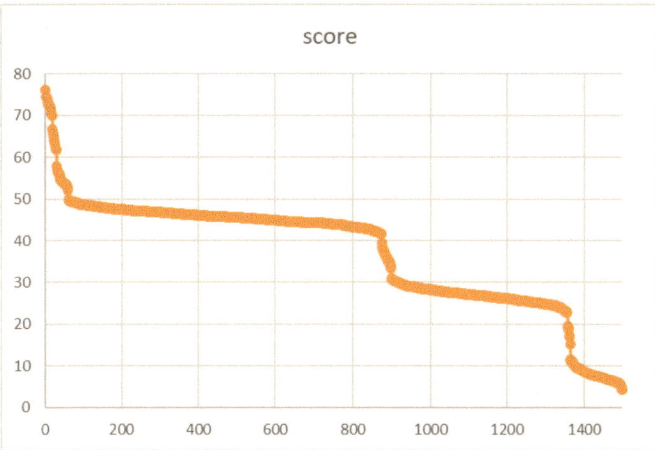

Figure 3.
Overall scores in descending order.

Score range	*Frequency*	*High dominance*	*Medium dominance*
71–80	17	2	2
61–70	12	0	1
51–60	31	2	8
41–50	814	46	115
31–40	39	0	5
21–30	442	29	63
11–20	19	2	1
1–10	126	4	24

Table 5.
Score-dominance map.

4. Conclusions

In this paper, we presented an MCDA approach for determining the eligibility of prisoners for early release. The criteria were set by the authorities based on vast experience and seemingly solid justifications. The values and weights of each criterion were defined in sessions that combined professional and academic insights.

It should be noted that the complexity of calculating the weights, the difficulty of maintaining internal traceability, and the inherent subjectivity of the process are valid concerns voiced by critics of the AHP process [19–21]. However, most of these pitfalls were avoided in the current analysis because the score matrix was objectively calculated using independent external resources. Moreover, the process was undertaken transparently and systematically to ensure fairness.

The method was implemented on a list of 1500 candidates for early release through a wide-ranging amnesty. Several hundred achieved overall scores that made them eligible for a pardon.

Acknowledgements

The author would like to thank the IDOD for the opportunity to assist in the discussed topic. The prisoners' data was intentionally left vague to respect the right to privacy. The author also would like to thank Mr. Yigal Walt for his editorial contribution.

Author details

Irit Talmor
Western Galilee College, Akko, Israel

*Address all correspondence to: iritt@wgalil.ac.il

References

[1] Larkin PJ, Clemency P. Good-time credits, and crowded prisons: Reconsidering early release. Georgetown Journal of Law Public Policy. Vol. 11. 2013. 44 pages

[2] Skarbek D. The Puzzle of Prison Order: Why Life behind Bars Varies around the World. Oxford University Press; 2020. p. 240

[3] Doumpos M, Zopounidis C. Multicriteria Analysis in Finance. Cham: Springer International Publishing; 2014. DOI: 10.1007/978-3-319-05864-1

[4] Liu H-C, Yang M, Zhou M, Tian G. An integrated multi-criteria decision making approach to location planning of electric vehicle charging stations. IEEE Transactions on Intelligent Transportation Systems. 2019;**20**: 362-373. DOI: 10.1109/ TITS.2018.2815680

[5] Wan C, Zhang D, Fang H. Incorporating AHP and evidential reasoning for quantitative evaluation of inland port performance. In: Lee PT-W, Yang Z, editors. Multi-Criteria Decision Making in Maritime Studies and Logistics: Applications and Cases. Cham: Springer International Publishing; 2018. pp. 151-173. DOI: 10.1007/978-3-319-62338-2_7

[6] Talmor I. Implementing a multi-criteria decision-making approach to a new party's election campaign – A case study. Methods X. 2021;**8**:101328. DOI: 10.1016/j.mex.2021.101328

[7] Kundakcı N. An integrated method using MACBETH and EDAS methods for evaluating steam boiler alternatives. Journal of Multi-Criteria Decision Analysis. 2019;**26**:27-34. DOI: 10.1002/ mcda.1656

[8] Triantaphyllou E. Multi-Criteria Decision Making Methods: A Comparative Study. US: Springer; 2000

[9] Ishizaka A, Nemery P. Multi-Criteria Decision Analysis. Chichester, UK: John Wiley & Sons Ltd; 2013. DOI: 10.1002/ 9781118644898

[10] Zopounidis C, Doumpos M, editors. Multiple Criteria Decision Making. 1st ed. Cham: Springer International Publishing; 2017. p. 211. DOI: 10. 1007/ 978-3-319-39292-9

[11] Saaty TL. A scaling method for priorities in hierarchical structures. Journal of Mathematical Psychology. 1977;**15**:234-281. DOI: 10.1016/ 0022-2496(77)90033-5

[12] Saaty TL. How to make a decision: The analytic hierarchy process. European Journal of Operational Research. 1990;**48**:9-26. DOI: 10.1016/ 0377-2217(90)90057-I

[13] Bonner HS, Rodriguez FA, Sorensen JR. Race, ethnicity, and prison disciplinary misconduct. Journal of Ethnicity in Criminal Justice. 2017;**15**:36-51

[14] Kurzfeld, J. Prison Crowding and Violent Misconduct (Internet). 2017. Available from: http://dx.doi.org/ 10.2139/ssrn.2994546 [Accessed 2023-02-06]

[15] Lahm KF. Inmate-on-inmate assault. Criminal Justice and Behavior. 2008;**35**: 120-137

[16] Hayes A, Burns A, Turnbull P, Shaw J. The health and social needs of older male prisoners. International Journal of Geriatric Psychiatry. 2012;**27**: 1155-1162

[17] Fazel S, Hope T, O'Donnell I, Piper M, Jacoby R. Health of elderly male prisoners: Worse than the general population, worse than younger prisoners. Age and Ageing. 2001;**30**: 403-407. DOI: 10.1093/ageing/30.5.403

[18] Klaus D. Goepel. AHP Priority Calculator n.d. Available from: https:// bpmsg.com/ahp/ahp-calc.php

[19] Asadabadi MR, Chang E, Saberi M. Are MCDM methods useful? A critical review of analytic hierarchy process (AHP) and analytic network process (ANP). Cogent Engineering. 2019;**6**: 1623153. DOI: 10.1080/23311916. 2019.1623153

[20] Noghin VD. What is the relative importance of criteria and how to use it in MCDM. In: Köksalan M, Zionts S, editors. Multiple Criteria Decision Making in the New Millennium. Berlin, Heidelberg: Springer Berlin Heidelberg; 2001. pp. 59-68

[21] Wątróbski J, Jankowski J, Ziemba P, Karczmarczyk A, Zioło M. Generalised framework for multi-criteria method selection. Omega. 2019;**86**:107-124. DOI: 10.1016/j.omega.2018.07.004

Chapter 4

Prioritization of the Physical Resilience Criteria for Affordable Housing Locating Based on An Analytic Hierarchy Process (AHP)

Mehrnaz Ramzanpour and Rouhollah Rahimi

Abstract

Natural hazards cause numerous problems occurred mostly in residential areas. Affordable housing is one of the types of housing that is planned for low- and middle-income groups. Optimum location of affordable housing is one of the most important criteria for this type of housing. Due to the high vulnerability of its residents, it is necessary to evaluate the site resilience. This chapter is aimed to identify and prioritize the criteria of physical resilience that are effective in selecting a resilient location for the affordable housing construction. Documentary materials are used to explain the literature and to determine the effective factors on physical resilience. Then the criteria were prioritized using AHP method by 22 experts. The important criteria obtained from the research include Infrastructure & Services (IS), Region Context (RC), Natural Environment (NE) and Surrounding Uses (SU). This research can be the basis of a strategic document for the discussion of the affordable housing resilience.

Keywords: physical resilience, affordable housing, locating, AHP, resilience criteria

1. Introduction

The location of housing and its connectivity to jobs, services, and amenities is central to reducing socio-spatial inequalities. Resilience is one of the most basic issues in today's world due to the occurrence of natural or anthropogenic events. The housing resilience process should have the necessary capability to predict accidents and prevent the burden of life and financial losses on citizens. In an ecosystem, resilience is a measure of the ecosystem's ability to absorb changes for its survival. Based on Holling's definition, resilience is determining the persistence of relations within a system and measuring the ability of this system to absorb changes created in various situations in order to resist various effects and factors. Based on the growth of urbanization and the increasing vulnerability of cities to disasters, significant changes have been observed in the attitude toward risks, as the dominant view has changed from focusing only on reducing vulnerability to increasing resilience against events. Risks around the world have always provided a great challenge in sustainable development, and as a result, some paths to achieve this development by reducing vulnerability patterns are essential and of great importance, and should find a suitable position in urban policies to create suitable conditions to reduce the effective and efficient risk at different levels. Thus, urban areas are considered sensitive areas due to the acceptance of large populations, which has increased the level of vulnerability due to the type of human interaction. Human being considers themselves a part of the place and based on their experiences of signs, meanings, and functions, he imagines a role for the location in his mind [1]. The identification and evaluation of such risks can be considered a good guide in the design of new urban areas and their control to manage urban areas by authorities.

Due to its climatic and geological characteristics and especially Iran's location on the Alpine-Himalaya earthquake belt, Iran is considered one of the most vulnerable countries around the world; so that the crisis risk index of the United Nations Development Program [2] indicates that, Iran has the highest vulnerability to earthquakes among the countries of the world after Armenia, and 31 of 40 types of natural disasters have taken place in Iran. Natural disasters have always been a great challenge for urban societies, and human settlements and infrastructures have always been endangered. According to statistics, accidents have increased over time and the urban communities' vulnerability to accidents (namely in developing countries) is increasing. Now, most of the world's population lives in cities. Major migrations from rural areas to cities and the formation of informal settlements, exposure of the population to unsuitable weather conditions, and other natural disasters are increasing and are great challenges for many cities. Under this condition, resilience, which can mean bouncing back from difficult events, becomes meaningful. In Iran, a country with a high probability of natural disasters and cities with heterogeneous and old textures, resilience is of great importance. One of the most important aspects of urban resilience is physical resilience.

The present study is aimed to measure the components of physical resilience in urban contexts for identifying resilient areas against natural disasters in order to build affordable housing. Affordable housing is a type of urban housing that is planned and built for groups with low- and middle-income socioeconomic levels with different policies. The optimal location of affordable housing is one of the most important planning criteria to build this type of housing [3]. Due to the high vulnerability of groups living in affordable housing, it is required to evaluate the resilience of its construction site in order to prevent the severity of the effect of natural and unnatural damages on this group in the future. On the other hand, presenting affordable housing is important to have a resilient city, and it is necessary to consider some criteria for it [4]. Studies have shown that the appropriate locating of social housing leads to city sustainability [5]. This research attempts to establish a relationship between affordable housing and physical resilience. This chapter is aimed to identify and prioritize the criteria of physical resilience that are effective in selecting a resilient location for the construction of affordable housing. Based on this goal, two questions of research are proposed:

- What are the criteria for resilient locating of affordable housing?
- How is the prioritization of these criteria?

This chapter has five main sections, first, the literature on resilience and resilience in the built environment is reviewed. The second part is related to the research method, in which the AHP model, study area, construction of questionnaire, and weighting of alternatives are done. In the third part, the results are stated, and in the fourth part, the discussion about the prioritization of the criteria and elements is discussed, and the limitations and suggestions of the research are stated. The last part is the conclusion.

2. Literature review

2.1 Resilience

The concept of resilience began in the 1960s and 1970s in the field of ecology. Then it was expanded to different fields of study and resilience became an interdisciplinary concept. The literature identifies two main approaches to resilience. The first approach is "equilibrium" related to the ability of an organism to resist shock and return to its original state, and the second approach is "evolutionary" or "transformational" related to the ability of an organism to continuously adapt to continuous changes in its environment [6]. The emergence of resilience as a driver of urban policy led to a shift toward an integrated and multidisciplinary planning system and quickly became an important urban policy discourse [7]. A city's ability to absorb disturbances while maintaining its functions and structures is a simple definition of resilience [8].

Resilience studies began in the late 1990s in response to environmental threats by adjusting social and institutional frameworks associated with planning, with a primary focus on physical and infrastructural improvements to prevent disruptions [9].

Resilience, which comes from the Latin root resi-lire meaning the return of spring, was first used by physical scientists. In the 1960s, with the rise of systems thinking, resilience entered the field of ecology, where multiple meanings of the concept have since emerged, each rooted in different worldviews and scientific traditions. The publication of an article in 1973 by a Canadian theoretical ecologist, Crawford Stanley Howling, advanced the concept of resilience, but although the concept of resilience has recently been added to the discourse of planners, it is by no means a new concept. Resilience researchers share a common understanding of resilience as a process involving change over time that produces a desirable outcome for one or more systems or parts of a system. For example, the dynamic interplay between resilience (change) and sustainability has been widely discussed for the first time by C.S. in a field of scientific studies that focuses on interactions between natural environments and human activities as social-ecological systems.

Holling [10] states that resilience is the "persistence of relationships within a system and a measure of the ability of these systems to absorb changes of state variables, driving variables, and parameters, and still persist."(p. 17) This idea has been extended by many researchers interested in international development studies and social-ecological systems. As Bousquet et al. [11] explain that resilience is "the capacity to cope with change and continue to develop," (p. 40) whether this development occurs for the survival of fisheries, forests, freshwater ecosystems, or communities dependent on these natural ecologies.

In other sciences, such as physical sciences, the same concept of stability and change is commonly used. For example, in architecture, the term resilience is synonymous with "the process of creating sustainable and successful places that promote well-being, by understanding people's needs for the places they live and work" [12]. The subject of resilience is also used in psychological sciences. Extensive studies have been conducted on promotional and protective processes. When human biological, psychological, social, economic, and political systems are stressed, the concept of resilience operates.

As a developmental psychologist, Masten [13] is known for evolving the concept of resilience to adopt a more systemic approach. He defines resilience as the capacity of a dynamic system to successfully adapt to disturbances that threaten system performance, viability, or development. This definition can be used for various systems at different interactive levels, both living and non-living, such as microorganisms, children, families, security systems, economy, forests, or global climate. Masten's definition has much in common with unrelated fields such as disaster resilience, where the focus is on "the ability to prepare and plan for, absorb, recover from, or adapt more successfully to actual or potential adverse events" [14].

Despite the distinction of all these definitions in the function of systems or different parts of systems, they have many similarities. First, resilience is associated with abnormal and stressful perturbations in one or more interdependent systems. Instability is an issue that threatens the system's capacity to maintain its performance. Second, all resilient systems involve processes to create opportunities for continuity, resilience, recovery, adaptation, or change.

Resilience is context-specific, as is the evolving public health thinking that currently emphasizes "precision public health" by identifying high-risk hotspots and then targeting interventions to their unique contexts. In this way, instead of generalization mechanisms, it seeks to maintain public welfare [15].

The third quality of resilience focuses on the need for sensitivity to the local context, acknowledging the different levels of power of each system (or part of a system) and its capacity to affect the individual or collective well-being of a system (or systems). Overall, this expression of power that leads to trade-offs is always a matter of debate as all the different parts of a system compete for the resources they need to deal with internal and external stressors.

Resilience is demonstrated only when a system functions in a way that is positively valued by its constituents or concurrent systems. In fact, it can be argued that a family that adopts criminal behavior as a way of managing social marginalization or economic adaptation to modernization in order to maintain its livelihood may be considered resilient from the perspective of those who benefit from these patterns [16].

Although the three definitions of resilience (i.e., exposure to an unusual disturbance, contextual specificity of protective processes, and negotiated outcomes) may seem abstract, in practice, resilience is a response to a disturbance that alters patterns of adaptation. Creates that favor some sectors. More than others, resilience has been the basis for a vast amount of study in many different disciplines. For example, Annarelli and Nonino [17] used Hollings' work on social-ecological systems to examine the resilience of supply chains. These include other environmental systems (for example, disruptive weather and political conflicts can be disruptive to supply chains) and the contemporary methods used by management (for example, labor strikes and poor financial decisions can affect the planned production of goods and services). Although the only desirable outcome of supply chain resilience may seem to be sustainable production (recovery), it must be said that a return to business as usual can be a very limited understanding of resilience.

2.2 Resilience in the built environment

The literature on resilience in the built environment can be expressed in three main paradigms: first, disciplines related to architecture, urban design, and planning. Second, combining ecosystem sciences with architecture and urban design. And third, resilience is embodied in architecture and built form. The first resilience paradigm is used as a model developed by architectural engineers, urban design, and planning.

The second resilience paradigm was shared as a theme between the disciplines of architecture and ecology. The third paradigm is also used as general knowledge individually and collectively through trial and error, learning, and memory. In the first paradigm, especially in areas with a high earthquake or hurricane risk, resilience thinking is used by combining contemporary practices of built environments with scientific approaches such as material physics and engineering resilience [18].

Hassler and Kohler [19] stated that the urban context is a complex socio-technical system with different time constants, actors, and institutional regimes that includes different scales such as buildings, neighborhoods, cities, and regions. They also used the term-built environment to refer to an artifact in an overlapping zone between culture and nature, with causality in both directions, referring to the relationship between the built and unbuilt parts of the environment. Pickett et al. [20] state that "resilient cities" are a metaphor for the integration of ecological, socio-economic, and planning realms. While metaphors have explanatory power for interdisciplinary discussions and can stimulate creativity around shared concepts and perspectives, metaphors are slippery figures of speech.

Resilience as an inclusive and open metaphor links the new non-equilibrium paradigm of ecosystem science with the dynamics of architecture, design, and urban planning, acknowledging that ecosystems are either externally regulated or have multiple or unstable state(s) or may have a dynamic state and possible disorder.

Belsky et al. [21] state that the trade-off that households make to reduce housing costs, such as transportation, access to public services, health and safety, is accounted for an ideal affordability assessment. Rowley [22] shows that in reality, housing affordability involves quality and location trade-offs. Monetary costs and socio-economic costs, which are hidden by traditional measures of affordability, are imposed on households by such trade-offs. It is clear that it is difficult and perhaps impossible to address all the concerns related to housing affordability, such as physical quality, location, access to services, and fit, in a simple measure, and they need to examine additional supplementary indicators [23].

3. Materials and methods

3.1 Development of the AHP model

The present study is based on the descriptive-analytical method. The documentary (library) method was used to explain the literature and records of the research subject and to determine the effective factors on physical resilience. Then, physical resilience indices were extracted for the location of affordable housing, based on theoretical justification. In this study, the combination of two important and highly applied mathematical models, i.e., fuzzy logic and hierarchical analysis process, has been used in order to rank indices. This ranking was performed using the AHP method with Expert choice v.11 software.

The AHP method was first presented by Saaty [24]. AHP is a multi-criteria decision-making method based on statistical data and a hierarchical structure. Normally, this method is designed to rank limited alternatives that have limited features. This method is used to eliminate probable errors for certain decision criteria. According to Saaty [25], the three main stages in the AHP approach include problem analysis, comparative analysis, and preference synthesis.

The components that best define the features of a resilient location for the construction of affordable housing are presented in a three-layer hierarchy of the AHP model (**Figure 1**), in which the top layer is the relevant goal, the second layer is the criteria, and the third layer is components.

The second layer consists of five physical resilience criteria, which include Surrounding Uses, Region Context, Natural Environment, Infrastructure & Services, and Open & Public Spaces. The elements of each measure of resilience are also the last layer.

This model was developed according to the literature review and expert approval. In this way, first, the resilience criteria were collected and categorized, then in order to coordinate with the issue of affordable housing, it was given to 10 architecture and urban planning experts (including university professors and doctoral students) for judgment. The criteria that all the experts approved were chosen as the final criteria of resilience in locating affordable housing. Since this kind of housing is built for society's low- and middle-income groups, its site should be chosen in such a way that it has physical resilience in critical conditions. So, it would be cost-effective and does not impose high costs on its residents in times of profound changes. On the other hand, it will be easier for these vulnerable residents to be stable and adapt to post-crisis conditions.

The criteria of physical resilience and the relevant components are obtained by evaluating the background and categorized in **Table 1**. According to Lanagarneshin et al. [29], the physical-environmental indices of resilience include green and open space, appropriateness of uses, characteristics of the land (ground), building

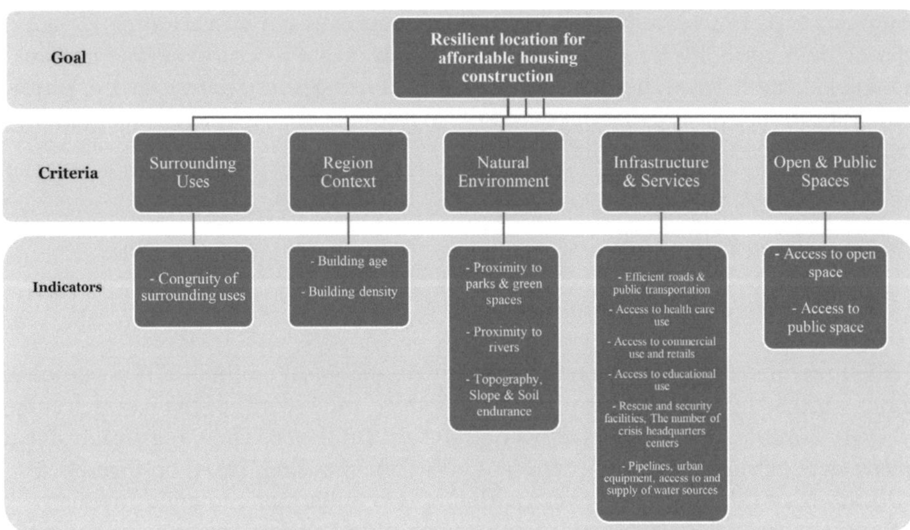

Figure 1.
AHP model used in the process of prioritizing criteria for resilient location for affordable housing.

Reference	Elements	Criteria
[26–30]	Congruity of surrounding uses	Surrounding Uses
[27, 31–33]	Buildings age (New, middle, worn, old)	Region Context
[26, 28–31, 33]	Building density	
[29–32, 34]	Proximity to parks & green spaces	Natural Environment
[28, 32, 34]	Proximity to superficial water (rivers, ...)	
[28, 29, 32, 34]	Topography, Slope & Soil endurance	
[26, 27, 29, 31–41]	Efficient roads & public transportation (accessibility routes)	Infrastructure & Services
[30–32, 34, 35, 37, 41, 42]	Access to health care use	
[32, 35, 42]	Access to commercial use and retails	
[37]	Access to educational use	
[27, 30, 31, 38]	Rescue and security facilities (Emergency, Police, Firefighting), The number of crisis headquarters centers	
[26–28, 32, 37, 38, 41]	Pipelines, urban equipment, access to and supply of water sources	
[26–30, 32]	Access to open space	Open & Public Spaces
[43–45]	Access to public space	

Table 1.
Physical resilience criteria and the relevant elements.

resistance, access, and density. Pashapour and Pourakrami [30] consider the physical dimensions of urban resilience as uses and characteristics of the surrounding context, access to urban facilities, rescue centers, parks, and open spaces. Delshad et al. [26] state that physical resilience against earthquakes can be measured with some criteria such as physical resistance, road network, infrastructural facilities and services, and the condition of open space. In their research, Farhadi et al. [27] introduced the physical measures affecting resilience by the uses surrounding the fabric, urban infrastructure including roads, crisis centers, and water supply, and access to open space. The dimensions of physical resilience against natural disasters have been presented by Ghasemi and Gharaee [28], resilience of place (feature of the land, distance from the river, etc.), resilience in the mental image (security ...), legal resilience (observing safety principles in constructions), structural resilience (building frame, roof lightness, etc.) and functional resilience (building density, width of roads, etc.). According to Desouza & Flanery [33], it is important to consider the characteristics of the fabric and access to roads and urban transportation in the design, planning, and management of resilient cities. Rezaie et al. [32] considered criteria in four structural-physical, socio-economic-cultural, spatial-functional, and structural-natural dimensions for analyzing the resilience of urban land use in Tehran. These criteria include the condition of buildings, the condition of relief and services, access network, employment centers, active population, population density, open and green space, worn texture, access to commercial uses, health and urban facilities and equipment, land bed and surface water. Abdullah et al. [46] evaluated the physical resilience in Tehran (Iran) with the indicators of population and building density, roads, worn-out texture, green and open spaces, health centers, crisis headquarters centers, and natural bed characteristics. Chen et al. [34] evaluated some indices such as topology, rate of green space, proximity to the river, building age, availability of medical facilities, and public services in the resilience of affordable housing. Regarding location resilience, Lyon [38] referred to the importance of infrastructural measures and public services during a crisis. In his research, Adger [35, 42] considers access to facilities such as roads, medical centers, and retail as important factors in resilience. In research on resilience in coastal areas, Orencio & Fujii [39] mentioned access to neighborhood facilities and road transportation as important in physical resilience. Pokhrel [40] evaluated green space in the resilience of the urban environment. He believed that access to roads for transportation is very important, especially during the crisis. Zhang et al. [41] consider urban resilience against climate change as dependent upon infrastructure criteria such as public services, transportation, and health use, access to infrastructure equipment, and crisis management. In their review research, Assarkhaniki et al. [37] considered the physical measures of resilience including access to road infrastructure, medical and educational centers, as well as urban infrastructure equipment and services during a crisis.

A brief description and definition of each criterion and elements are given in **Table 2**.

Description	Elements	Description	Criteria
Coordination of surrounding uses with the site	Congruity of surrounding uses	Parcel of land with different use around [a]the site	Surrounding Uses
Quality of the building based on the construction time	Buildings age (New, middle, worn, old)	Surrounding building conditions in the area	Region Context
Ratio of building structures per acre	Building density		
Number of parks & green spaces in the vicinity of the site	Proximity to parks & green spaces	Non-human-made surroundings and conditions	Natural Environment
Number of rivers or other superficial water in the vicinity of the site	Proximity to superficial water (rivers, ...)		
The chosen land (site) physical conditions	Topography, Slope & Soil endurance		
Efficient connectivity to city areas with appropriate roads and transportation	Efficient roads & public transportation (accessibility routes)	City infrastructures and facilities that are accessible for the site	Infrastructure & Services
Site proximity to health care use (clinic, hospital, ...)	Access to health care use		
Site proximity to commercial use and retails (mall, grocery, ...)	Access to commercial use and retails		

Description	Elements	Description	Criteria
Site proximity to schools, universities ...,	Access to educational use		
Site proximity to Rescue and security facilities and crisis headquarters centers	Rescue and security facilities (Emergency, Police, Firefighting), The number of crisis headquarters centers		
Site accessibility to some equipment	Pipelines, urban equipment, access to and supply of water sources		
Site proximity to open spaces among outside buildings	Access to open space	Any urban ground space with unrestricted access	Open & Public Spaces
Site proximity places of public use (park, square, ...)	Access to public space		

[a] *Affordable housing site.*

Table 2.
Description of criteria/elements.

3.2 Reliability and validity

Before the conduction of the survey on the developers, validation, assessment, and rationality of the questionnaire was carried out. This was achieved with the aid of practitioners and studies lecturers; they evaluated the questionnaire in order to put off the anomaly of expression and make sure relevant terms are used primarily based on the peculiarity of physical resilience. The questionnaire became finalized based totally on the form of comments. Information reliability testing revealed a Cronbach alpha rate of 0.79 for the 24 identified elements. This is a little greater than the boundary of 0.70. The collected records are remarkably dependable for further statistical evaluation.

3.3 Study area and local decision-makers

The prioritizing process of the resilience components of affordable housing was performed in Mazandaran province (**Figure 2**). The familiarity and experience of these specialists with Mazandaran and the resilience resources in its cities was the main reason for selecting them. These members from the academic environment and local government (with specialized responsibilities in this field) were selected as decision-makers during the prioritization of resilience components due to their experience, skills, knowledge, and activities related to the topic. The questionnaire was distributed among 22 experts (See Appendix 1 and 2). Experts' profiles are shown in **Table 3**.

Lam and Zhao [47] argued that AHP survey is an unusual approach for research associated with a particular trouble; subsequently the adoption of a massive sample is not always imperative. Tavares et al. [48] argued that the peculiarity of AHP makes judgment from one professional be deemed ok. On the contrary, Cheng and Li [49]

Figure 2.
Map of the study area, Mazandaran province, Iran.

Variable	Type		Distribution
Role	Academic	Ph.D. Candidate	5 (22.7%)
		Ph.D. (University Prof.)	10 (45.5%)
		Post-doc	1 (4.5%)
	Local government		6 (27.3%)
Age	30–40		6 (27.3%)
	41–50		9 (40.9%)
	Up to 51		7 (31.8%)
Gender	Male		13 (59%)
	Female		9 (41%)
Years of experience	5–10		7 (32%)
	Up to 10		15 (68%)

Table 3.
Profiles of the participants in the analytic hierarchy process (AHP) study.

advised that the adoption of a big sample size for an AHP examine can also result in inconsistent judgment, as many professionals may additionally provide arbitrary consequences. The peculiarity of AHP in housing research and construction studies might be tied to its functionality to cope with small pattern sizes. Research [50–52] has followed respondents starting from four to nine even as others used a pattern size of 20 to 30 [53, 54]. As most of the people of the research followed a small sample size, it is imperative that to allow beneficial decisions, regular outputs and models, the adoption of a small pattern size is most desirable. Therefore 22 developers from the overall survey with over 6 years' experience in sustainable housing were selected to take part within the AHP survey.

3.4 Weights of alternatives in a consistent matrix

The paired comparisons are a key step in AHP to prioritize the weights of locating factors. Thus, based on different factors, the alternatives are scored. This process simultaneously focuses on two factors and their relations with each other. The relative importance of each factor is measured via a numerical measurement scale [55]. According to the AHP model, the identification of important criteria and components of resilience in affordable housing locating was performed using paired comparisons and ratio-scale measurement, which is explained as n(n-1)/2, where n denotes the components in a prioritization judgment [56]. In this research, 10 comparisons in a matrix are defined for 5 criteria, while the comparison of criteria components formed 90 questions. Each output of a paired comparison shows the decision-maker's preferences for one alternative over the others based on a set of scales (**Table 4**) that include scales from 1 to 9 [24, 56].

When alternative *i* was considered extremely important compared with alternative *j*, the calculation matrix score was based on *aij* = 9 and *aji* = 1/9. The distribution of these scores in a square matrix resulted in a reciprocal matrix [57], represented as:

$$A = [a_{ij}] = \begin{Bmatrix} 1 & a_{ij} \ldots & a_{1n} \\ 1/a_{ij} & 1 \ldots & a_{2n} \\ \vdots & \vdots & \vdots \\ 1/a_{1n} & 1/a_{2n} \ldots & 1 \end{Bmatrix} \tag{1}$$

where A = [*aij*] is a representation of the intensity of the decision-maker's preference for one over another compared alternative *aij* and for all comparisons *i*, *j* = 1,2, ... n. Decision-makers facilitated the comparisons of alternative criteria or elements in rounds till the ratings had been taken into consideration strong. Stability was reached while a certain consensus on a sum of rankings became done [39].

Consistency test: This test is conducted to check the judgment consistency. This test guarantees that best steady matrixes are covered in similarly evaluation. The formulation utilized in calculating the best eigenvalue and vector is:

Scale	Judgment of preference	Description
1	Equally important	Two factors contribute equally to the objective
3	Moderately important	Experience and judgment slightly favor one over the other
5	Strongly important	Experience and judgment strongly favor one over the other
7	very strongly important	Experience and judgment very strongly favor one over the other, as demonstrated in practice
9	Extremely important	The evidence favoring one over the other is of the highest possible validity
2.4. 6, 8	Intermediate preferences between adjacent scales	When compromise is needed

Table 4.
Rating scale for judging preferences used for the pair wise comparison of various physical resilience criteria in affordable housing locating.

$$\lambda_{max} = \sum_{j=1}^{m} \frac{Aw}{Mw1} (i = 1, 2, \dots, m) \quad (2)$$

Where λmax represents matrix the highest eigenvalue, A denotes the pairwise matrix; w stands for the matrix of weights of elements, and wi stands for the element's weights. The consistent level of the judgment was determined using the Consistency Ratio (CR) which was computed by the formula:

$$\text{CR} = \frac{CI}{RI} = \frac{1}{RI} \left(\frac{\lambda_{max} - m}{m - 1} \right) \quad (3)$$

Here, CI uses an integrity index. CR is the consistency ratio. RI represents the random index and m represents the amount of CSF in the matrix.

Weight calculation:

CSF weights were determined by estimating the eigenvector matrix and a measure of the consistency of judgment is obtained by computing the maximum eigenvalue. In the AHP study, we calculated CSF weights at each stratum level to establish priority among elements. This was acquired through:

$$n_i = \prod_{j=1}^{m} aij (i = 1, 2, \dots, m) \quad (4)$$

Where ni represents the multiplication of the relative importance for each row of CSFs; aij represents the relative importance of the CSFs "i" were compared with CSFs "j" and m represents the number of CSFs in the matrix. Vector wi was calculated by:

$$w_i = \sqrt[m]{\overline{n}_i (i = (1, 2, \dots, m)} \quad (5)$$

Where wi represents the mth power of ni:

$$w_i = \frac{w_i}{\sum_{i=1}^{m} w_i} (i = \overline{1, \underline{2,}} \dots, m) \quad (6)$$

The weights of the CSFs were calculated by normalization of the vector: where wi stands for weights of CSFs and criteria.

3.5 Consensus building

To acquire consensus on ratings in a paired comparison of variables in the AHP model, the Delphi method became selected as a powerful technique in a multilateral decision-making procedure. But the technique needed a robust facilitator that would coordinate the distinct views of the decision-makers toward a single objective. Regardless of comparable experiences, social status (e.g., education), and engagement level with the issue of resilience among respondents, the results confirmed differences in perspectives about each variable.

The Delphi method was of great importance during the comparison of the variables and the level of each one. Decision-makers tended to consider variables as having similar goals, which made comparisons difficult. The facilitator's role was to interpret the differences between the variables and to organize the respondents' views. Thus, the group managed to discover a common perspective for the superiority of each variable in paired comparisons.

The use of a quantitative base scale (**Table 2**) helped the respondents in scoring each pairwise comparison, especially when the number of variables of each criterion was above two. To make scoring easier, the variables were placed in the table in front of each other and the scores (1 to 9 on each side) were between them.

The agreement on the final scores was obtained from a pairwise comparison of criteria and components using Delphi technique. Final scores were calculated based on the geometric mean of all scores given by the decision-makers for each pairwise comparison. When consensus was achieved, a summary of the final scores for each pairwise comparison was entered into a matrix or decision table.

4. Results

4.1 Selected criteria and elements

The comparison matrix at the criterion level was consistent with a value of 0.08 (**Table 5**). Based on the weights of alternatives at this level, Infrastructure & Services (IS) and Surrounding Uses (SU) were ranked as the highest and lowest criteria, respectively. The highest ranked criteria, i.e., Infrastructure & Services (IS), Region Context (RC), and Natural Environment (NE), were selected by the sum of their weights and accounted for 85% of the overall weights of the criteria being compared. Their attribute elements were then subjected to further comparison, and high-ranking elements were subsequently selected.

For Infrastructure & Services (IS), the elements that characterized resilient location for affordable housing were IS1, IS2, IS3, IS4, IS5, and IS6 which accounted for 75% of the overall alternatives (**Table 6**). Among its elements, IS5 accounted for 30% of the most important attributes that describe the resilient location. The matrices of comparisons for these attribute elements fell within a CR value of 0.10 and 0.09, respectively.

Subsequent procedures for selecting and evaluating attribute elements were conducted for Region Context (RC), and Natural Environment (NE). For Region Context (RC), the elements RC1, and RC2 were selected as elements that describe the resilient location of affordable housing, in which, RC2 is a more important element (**Table 6**). For Natural Environment (NE), the elements NE1, NE2, and NE3 were selected to represent elements that described the resilient location of affordable housing, in which, NE3 is the most important element.

Codes	Criteria	Weight	Rank
SU	Surrounding Uses	.052	5
RC	Region Context	.184	2
NE	Natural Environment	.102	3
IS	Infrastructure & Services	.564	1
OPS	Open & Public Spaces	.098	4

Table 5.
Normalized Weights and ranks of various criteria of a resilient location of affordable housing.

Criteria		Elements of resilient location of affordable housing	Weights	Rank
SU	SU1	Congruity of surrounding uses	.021	11
RC	RC1	Buildings age	.025	10
	RC2	Building density	.076	5
NE	NE1	Parks & green spaces	.006	14
	NE2	Proximity to rivers	.042	6
	NE3	Topography & Slope & Soil endurance	.28	9
IS	IS1	Efficient roads & public transportation	.164	3
	IS2	Access to health care use	.182	2
	IS3	Access to commercial use and retails	.033	8
	IS4	Access to educational use	.018	12
	IS5	Rescue and security facilities, the number of crisis headquarters centers	.233	1
	IS6	Pipelines, urban equipment, access to and supply of water sources	.123	4
OPS	OPS1	Access to open space	.040	7
	OPS2	Access to public space	.007	13

Table 6.
Normalized Weights and ranks of various elements that characterized the selected criteria to produce a resilient location of affordable housing.

5. Discussion

5.1 Priority criteria and elements

Infrastructure & Services (IS) was the most important criterion for finding a physically resilient location for setting the affordable housing because the existence of some infrastructure is crucial to the resilience of residential projects.

Among the variables of service and infrastructure, rescue and security facilities, access to health care use, efficient roads & public transportation, and urban equipment were the most important, respectively. Urban—water infrastructure, sewage, energy, communication and transportation systems—are of great importance in emergency response and rapid recovery of the community and its economy. Critical systems should be designed so they are not destroyed by natural risks. The transportation infrastructure network and the structural reliability of the network components (roads and bridges) are the most important measures of physical resilience [26, 58–60]. Proximity to suitable rescue uses includes crisis management centers and health uses. The logical location of the main components of the city and the logical relationships between them help to create a resilient region. On the other hand, the damage to infrastructural facilities such as water, electricity, and gas networks can greatly increase the damages caused by earthquakes in the city [26]. Some of the utilities in the city play a great role in the city's vulnerability to natural disasters. These uses are known as special uses including schools, universities, hospitals, rescue centers, urban management centers, factories, and fuel tanks [30].

Region Context (RC) represents the physical condition of context that has surrounded an affordable housing project. The degree of security of discrete context against natural disasters is higher than the degree of security of continuous context. The more regular the segmentation pattern and the less obtuse angles, the less the degree of vulnerability. Section area, section length, and width proportions in relation to land use and ownership type will be effective in the vulnerability factor or context efficiency [61]. If there are worn-out contexts in the area, strengthening, restoring, restoration, and renovating these old contexts seem necessary. Due to its direct relationship with population density, the density of living collections indicates the number of financial losses and life losses in the earthquake and the increase of the crisis. The relationship between population density and earthquake impacts is complex. Based on the inductive and reasoning method, it is obvious that population density is not effective on the intensity of destruction, but the importance of densities is related to the occurrence of destruction [62]. As the building gets older, the resilience is reduced.

Natural Environment (NE) describes natural resources in the vicinity of affordable housing project sites which affects the resilience of the location. It reflects the degree of vulnerability to flooding, waterlogging, and proximity to the river in the affordable housing area. The topography is characterized by the difference between the highest topography and the lowest topography within the community. Topography and land slope are important factors in guiding surface water in case of danger [34].

Open & Public Spaces (OPS) and Surrounding Uses (SR), are respectively the least important criteria at the time of finding a resilient location for affordable housing construction. Access to suitable open spaces to escape from dangerous factors and access to safe places, the possibility of quick and safe escape and shelter, facilitating relief and rescue operations after the earthquake, speeding up cleaning operations are essential. Therefore, it is the degree and quality of open spaces that can act as the additional element of destroyed buildings and infrastructures in the event of an earthquake crisis, therefore, open spaces play a crucial role in the physical resilience of space against earthquakes [26, 27, 32]. If the ratio of green space is high, the regulation ability of rainwater infiltration is also good [34]. Moreover, green spaces are important components of the urban landscape and improve the physical, psychological stress relief center and improve social harmony through social interaction and recreation. Further, these green spaces help to protect urban biodiversity to protect urban landscape through mitigation and adaptation of adverse impacts of climate change, through heat reduction and cooling effects. In the same way, the green space also helps to maintain the hydrology around the city, provide emergency shelter, and many more [40].

5.2 Limitations of the proposed criteria and future research

The result of this research demonstrates that the AHP model can provide a framework to assist decision-makers in analyzing various location factors, evaluating location site alternatives, and making final location selections based on resilience. One of the limitations of the research is the sampling of decision-makers. Due to choosing the north of Iran as the study area, experts from this region of the country were selected to respond to the AHP questionnaire. It is required to perform a survey in other geographical locations with larger and different groups of experts.

In this research, only the discovery and prioritization of resilience criteria were discussed. Thus, it is necessary to evaluate these prioritized criteria in the real environment to select a suitable site for the construction of affordable housing or to evaluate the resilience of existing affordable housing projects in future research. Social criteria are also the factors affecting the resilience of affordable housing. Hence, it is necessary to examine the criteria of social resilience and their relationship with the physical criteria of resilience in selecting an affordable housing site. Future research extensions could also investigate the impact of changing input parameters such as measurements and the importance of location factors. Further studies of sensitivity analyses on the effects of changes in decision-maker preferences are needed, as changes in decision-maker preferences can affect the attractiveness of a particular place.

6. Conclusions

The issue of housing is one of the most important parts of development in a society, which has a great impact on the health and view of the society, with its great economic, social, environmental, and physical dimensions. Today, despite many advances, problems in the housing sector are one of the challenging issues of developing countries, such as internal migrations, problems of land supply, inadequate resources, weak economic management, lack of comprehensive housing planning, and other inefficiencies which exist in the economic infrastructure of these countries, on the one hand, and the rapid increase of the urban population, on the other hand, has turned to present shelter in these countries into an unsolved problem. Above all, the low quality of urban housing in the cities has led to the vulnerability of housing and their lack of resilience against all kinds of accidents and based on the available statistics, there are many casualties of life and loss of citizens due to the low quality of buildings and lack of resilience. Evaluating housing indices is one of the various recognized tools and methods of housing characteristics, which can be used to identify the effective procedures of housing. Based on the low percentage of durable houses in Iran, the need to evaluate resilience in physical indices is intensified, especially in times of crisis. Due to the strong role of cities in the loss of lives and financial aspects of citizens, urban crisis management theories with emphasis on making the city resilient and especially the resilience of urban houses have turned into an important strategy for cities in less than a decade, because the housing sector is of great importance as one of the great topics in this strategy.

In this chapter, by extracting the physical components of resilience and measuring the opinions of experts, the prioritization of the components and indices of physical resilience in urban areas for the location of affordable housing was considered. In order to respond to the research questions, the components affecting physical resilience are services and infrastructures, the context of the region, the natural environment, open and public spaces, and the surrounding uses, respectively. In the services and infrastructure components, the highest impact factor is related to access to rescue and security facilities, the number of crisis headquarters centers and smart infrastructures, and access to health facilities. In terms of the effect of physical indicators on the resilience of urban areas, research results are consistent with research results [26, 27, 30, 32, 58–61]. The important point of this research compared to previous research is that it has focused on identifying the most resilient urban land to build affordable housing by identifying and prioritizing the physical components of resilience in those areas with the neighboring context which can be the basis of a strategic document for the location of affordable housing lands at the macro level. Creating resilience needs cooperation and communication between organizations and stakeholders, adapting the management institution to the ecological scale of the resource, and preventing section-based analysis. It is rarely possible to find and even build a city that completely has resilience components and indices, but what is important is the will of these cities and their urban management and their step-by-step movement toward prepared cities and getting closer to resilient cities. In order to achieve this purpose, the campaign to construct resilient cities is presenting guidance and helps city managers evaluate the current situation of cities based on the approved standards of prepared and resilient cities, and also it attempts to parallel the growth and development of cities and moving them on the path of resilient cities.

Acknowledgements

This research work has been supported by a research grant from the University of Mazandaran.

Conflict of interest

"The authors declare no conflict of interest."

A. Appendix 1. AHP questionnaire 1

Dear respondent: Thank you for taking the time to complete this questionnaire. The following questionnaire has been prepared for "locating affordable housing with approach of physical resilience." Please complete it. Thank you for your sincere cooperation.

Age: ...

Gender: Male ☐ Female ☐

Field of study: Architectural Engineering ☐ Urban Engineering ☐

Education level: Ph.D. Candidate ☐ Ph.D. ☐ Post-doc ☐

Role: ...

* Description: In this questionnaire, please, as an expert, determine which criterion and to what extent do you think is more important for locating low-income housing. Criteria and sub-criteria in this study are:

1. Surrounding Uses (Congruity of surrounding use).

2. Region Context (Buildings age (New, middle, worn, old); Building density).

3. Natural Environment (Proximity to parks & green spaces; Proximity to superficial water (rivers, ...), Topography; Slope & Soil endurance).

4. Infrastructure & Services (Efficient roads & public transportation (accessibility routes); Access to health care use; Access to commercial use and retails; Access to educational use; Rescue and security facilities (Emergency, Police, Firefighting); The number of crisis headquarters centers; Pipelines, urban equipment, access to and supply of water sources).

5. Open & Public Spaces (Access to open space; Access to public space).

* Please compare these criteria to find out which one is more important than the other and how important it is. Please rank between 1 and 9.

1- How important is the Surrounding Uses factor compared to the Region Context factor for the resilient location of affordable housing?

Region Context					Surrounding Uses			
Absolutely important	Very important	More important	Slightly important	Equally important	Slightly important	More important	Very important	Absolutely important
☐	☐	☐	☐	☐	☐	☐	☐	☐

2- How important is the Surrounding Uses factor compared to the Natural Environment factor for the resilient location of affordable housing?

Natural Environment					Surrounding Uses			
Absolutely important	Very important	More important	Slightly important	Equally important	Slightly important	More important	Very important	Absolutely important
								☐

3- How important is the Surrounding Uses factor compared to the Infrastructure & Services factor for the resilient location of affordable housing?

Infrastructure & Services					Surrounding Uses			
Absolutely important	Very important	More important	Slightly important	Equally important	Slightly important	More important	Very important	Absolutely important
☐	☐	☐	☐	☐	☐	☐	☐	☐

4- How important is the Surrounding Uses factor compared to the Open & Public Spaces factor for the resilient location of affordable housing?

Open & Public Spaces					Surrounding Uses			
Absolutely important	Very important	More important	Slightly important	Equally important	Slightly important	More important	Very important	Absolutely important
☐	☐	☐	☐	☐	☐	☐	☐	☐

5- How important is the Region Context factor compared to the Natural Environment factor for the resilient location of affordable housing?

Natural Environment					Region Context			
Absolutely important	Very important	More important	Slightly important	Equally important	Slightly important	More important	Very important	Absolutely important
☐	☐	☐	☐	☐	☐	☐	☐	☐

6- How important is the Region Context factor compared to the Infrastructure & Services factor for the resilient location of affordable housing?

Infrastructure & Services — Region Context

Absolutely important	Very important	More important	Slightly important	Equally important	Slightly important	More important	Very important	Absolutely important
☐	☐	☐	☐	☐	☐	☐	☐	☐

7- How important is the Region Context factor compared to the Open & Public Spaces factor for the resilient location of affordable housing?

Open & Public Spaces					Region Context			
Absolutely important	Very important	More important	Slightly important	Equally important	Slightly important	More important	Very important	Absolutely important
☐	☐	☐	☐	☐	☐	☐	☐	☐

8- How important is the Natural Environment factor compared to the Infrastructure & Services factor for the resilient location of affordable housing?

Infrastructure & Services					Natural Environment			
Absolutely important	Very important	More important	Slightly important	Equally important	Slightly important	More important	Very important	Absolutely important
☐	☐	☐	☐	☐	☐	☐	☐	☐

9- How important is the Natural Environment factor compared to the Open & Public Spaces factor for the resilient location of affordable housing?

Open & Public Spaces					Natural Environment			
Absolutely important	Very important	More important	Slightly important	Equally important	Slightly important	More important	Very important	Absolutely important
☐	☐	☐	☐	☐	☐	☐	☐	☐

10- How important is the Infrastructure & Services factor compared to the Open & Public Spaces factor for the resilient location of affordable housing?

Open & Public Spaces					Infrastructure & Services			
Absolutely important	Very important	More important	Slightly important	Equally important	Slightly important	More important	Very important	Absolutely important
☐	☐	☐	☐	☐	☐	☐	☐	☐

B. Appendix 2. AHP questionnaire 2.

Dear respondent: Thank you for taking the time to complete this questionnaire. The following questionnaire has been prepared for "locating affordable housing with approach of physical resilience." Please complete it. Thank you for your sincere cooperation.

Age: ...

Gender: Male ☐ Female ☐.

Field of study: Architectural Engineering ☐ Urban Engineering ☐.

Education level: Ph.D. Candidate ☐ Ph.D. ☐ Post-doc ☐.

Role: ...

* Description: In the comparison of criterion j with criterion i, if the importance of both criteria is the same, mark the number 1. If the criterion on the right was more important, the number on the right should be chosen as much as it is more important. But if the number on the left is more important, mark the number on the left. Be careful to only mark the number of one side (the side of the item that is more important).

Sub-criteria (elements)	Priorities	Sub-criteria (elements)
Buildings age (New, middle, worn, old)	9 8 7 6 5 4 3 2 1 2 3 4 5 6 7 8 9	Congruity of surrounding uses
Building density	9 8 7 6 5 4 3 2 1 2 3 4 5 6 7 8 9	Congruity of surrounding uses
Proximity to parks & green spaces	9 8 7 6 5 4 3 2 1 2 3 4 5 6 7 8 9	Congruity of surrounding uses
Proximity to superficial water (rivers, ...)	9 8 7 6 5 4 3 2 1 2 3 4 5 6 7 8 9	Congruity of surrounding uses
Topography, Slope & Soil endurance	9 8 7 6 5 4 3 2 1 2 3 4 5 6 7 8 9	Congruity of surrounding uses
Efficient roads & public transportation (accessibility routes)	9 8 7 6 5 4 3 2 1 2 3 4 5 6 7 8 9	Congruity of surrounding uses
Access to health care use	9 8 7 6 5 4 3 2 1 2 3 4 5 6 7 8 9	Congruity of surrounding uses
Access to commercial use and retails	9 8 7 6 5 4 3 2 1 2 3 4 5 6 7 8 9	Congruity of surrounding uses
Access to educational use	9 8 7 6 5 4 3 2 1 2 3 4 5 6 7 8 9	Congruity of surrounding uses
Rescue and security facilities, the number of crisis headquarters centers	9 8 7 6 5 4 3 2 1 2 3 4 5 6 7 8 9	Congruity of surrounding uses
Pipelines, urban equipment, access to and supply of water sources	9 8 7 6 5 4 3 2 1 2 3 4 5 6 7 8 9	Congruity of surrounding uses
Access to open space	9 8 7 6 5 4 3 2 1 2 3 4 5 6 7 8 9	Congruity of surrounding uses
Access to public space	9 8 7 6 5 4 3 2 1 2 3 4 5 6 7 8 9	Congruity of surrounding uses

Building density	9 8 7 6 5 4 3 2 1 2 3 4 5 6 7 8 9	Buildings age (New, middle, worn, old)
Proximity to parks & green spaces	9 8 7 6 5 4 3 2 1 2 3 4 5 6 7 8 9	Buildings age
Proximity to superficial water (rivers, ...)	9 8 7 6 5 4 3 2 1 2 3 4 5 6 7 8 9	Buildings age
Topography, Slope & Soil endurance	9 8 7 6 5 4 3 2 1 2 3 4 5 6 7 8 9	Buildings age
Efficient roads & public transportation (accessibility routes)	9 8 7 6 5 4 3 2 1 2 3 4 5 6 7 8 9	Buildings age

Sub-criteria (elements)	Priorities																	Sub-criteria (elements)
Access to health care use	9	8	7	6	5	4	3	2	1	2	3	4	5	6	7	8	9	Buildings age
Access to commercial use and retails	9	8	7	6	5	4	3	2	1	2	3	4	5	6	7	8	9	Buildings age
Access to educational use	9	8	7	6	5	4	3	2	1	2	3	4	5	6	7	8	9	Buildings age
Rescue and security facilities, the number of crisis headquarters centers	9	8	7	6	5	4	3	2	1	2	3	4	5	6	7	8	9	Buildings age
Pipelines, urban equipment, access to and supply of water sources	9	8	7	6	5	4	3	2	1	2	3	4	5	6	7	8	9	Buildings age
Access to open space	9	8	7	6	5	4	3	2	1	2	3	4	5	6	7	8	9	Buildings age
Access to public space	9	8	7	6	5	4	3	2	1	2	3	4	5	6	7	8	9	Buildings age
Proximity to parks & green spaces	9	8	7	6	5	4	3	2	1	2	3	4	5	6	7	8	9	Building density
Proximity to superficial water (rivers, ...)	9	8	7	6	5	4	3	2	1	2	3	4	5	6	7	8	9	Building density
Topography, Slope & Soil endurance	9	8	7	6	5	4	3	2	1	2	3	4	5	6	7	8	9	Building density
Efficient roads & public transportation (accessibility routes)	9	8	7	6	5	4	3	2	1	2	3	4	5	6	7	8	9	Building density
Access to health care use	9	8	7	6	5	4	3	2	1	2	3	4	5	6	7	8	9	Building density
Access to commercial use and retails	9	8	7	6	5	4	3	2	1	2	3	4	5	6	7	8	9	Building density
Access to educational use	9	8	7	6	5	4	3	2	1	2	3	4	5	6	7	8	9	Building density
Rescue and security facilities, the number of crisis headquarters centers	9	8	7	6	5	4	3	2	1	2	3	4	5	6	7	8	9	Building density
Pipelines, urban equipment, access to and supply of water sources	9	8	7	6	5	4	3	2	1	2	3	4	5	6	7	8	9	Building density
Access to open space	9	8	7	6	5	4	3	2	1	2	3	4	5	6	7	8	9	Building density
Access to public space	9	8	7	6	5	4	3	2	1	2	3	4	5	6	7	8	9	Building density

Sub-criteria (elements)	Priorities	Sub-criteria (elements)
Proximity to superficial water (rivers, ...)	9 8 7 6 5 4 3 2 1 2 3 4 5 6 7 8 9	Proximity to parks & green spaces
Topography, Slope & Soil endurance	9 8 7 6 5 4 3 2 1 2 3 4 5 6 7 8 9	Proximity to parks & green spaces
Efficient roads & public transportation (accessibility routes)	9 8 7 6 5 4 3 2 1 2 3 4 5 6 7 8 9	Proximity to parks & green spaces
Access to health care use	9 8 7 6 5 4 3 2 1 2 3 4 5 6 7 8 9	Proximity to parks & green spaces
Access to commercial use and retails	9 8 7 6 5 4 3 2 1 2 3 4 5 6 7 8 9	Proximity to parks & green spaces
Access to educational use	9 8 7 6 5 4 3 2 1 2 3 4 5 6 7 8 9	Proximity to parks & green spaces
Rescue and security facilities, the number of crisis headquarters centers	9 8 7 6 5 4 3 2 1 2 3 4 5 6 7 8 9	Proximity to parks & green spaces
Pipelines, urban equipment, access to and supply of water sources	9 8 7 6 5 4 3 2 1 2 3 4 5 6 7 8 9	Proximity to parks & green spaces
Access to open space	9 8 7 6 5 4 3 2 1 2 3 4 5 6 7 8 9	Proximity to parks & green spaces
Access to public space	9 8 7 6 5 4 3 2 1 2 3 4 5 6 7 8 9	Proximity to parks & green spaces
Topography, Slope & Soil endurance	9 8 7 6 5 4 3 2 1 2 3 4 5 6 7 8 9	Proximity to superficial water (rivers, ...)
Efficient roads & public transportation (accessibility routes)	9 8 7 6 5 4 3 2 1 2 3 4 5 6 7 8 9	Proximity to superficial water (rivers, ...)
Access to health care use	9 8 7 6 5 4 3 2 1 2 3 4 5 6 7 8 9	Proximity to superficial water (rivers, ...)
Access to commercial use and retails	9 8 7 6 5 4 3 2 1 2 3 4 5 6 7 8 9	Proximity to superficial water (rivers, ...)
Access to educational use	9 8 7 6 5 4 3 2 1 2 3 4 5 6 7 8 9	Proximity to superficial water (rivers, ...)
Rescue and security facilities, the number of crisis headquarters centers	9 8 7 6 5 4 3 2 1 2 3 4 5 6 7 8 9	Proximity to superficial water (rivers, ...)
Pipelines, urban equipment, access to	9 8 7 6 5 4 3 2 1 2 3 4 5 6 7 8 9	Proximity to superficial water (rivers, ...)

Sub-criteria (elements)	Priorities	Sub-criteria (elements)
and supply of water sources		
Access to open space	9 8 7 6 5 4 3 2 1 2 3 4 5 6 7 8 9	Proximity to superficial water (rivers, ...)
Access to public space	9 8 7 6 5 4 3 2 1 2 3 4 5 6 7 8 9	Proximity to superficial water (rivers, ...)
Efficient roads & public transportation (accessibility routes)	9 8 7 6 5 4 3 2 1 2 3 4 5 6 7 8 9	Topography, Slope & Soil endurance
Access to health care use	9 8 7 6 5 4 3 2 1 2 3 4 5 6 7 8 9	Topography, Slope & Soil endurance
Access to commercial use and retails	9 8 7 6 5 4 3 2 1 2 3 4 5 6 7 8 9	Topography, Slope & Soil endurance
Access to educational use	9 8 7 6 5 4 3 2 1 2 3 4 5 6 7 8 9	Topography, Slope & Soil endurance
Rescue and security facilities, the number of crisis headquarters centers	9 8 7 6 5 4 3 2 1 2 3 4 5 6 7 8 9	Topography, Slope & Soil endurance
Pipelines, urban equipment, access to and supply of water sources	9 8 7 6 5 4 3 2 1 2 3 4 5 6 7 8 9	Topography, Slope & Soil endurance
Access to open space	9 8 7 6 5 4 3 2 1 2 3 4 5 6 7 8 9	Topography, Slope & Soil endurance
Access to public space	9 8 7 6 5 4 3 2 1 2 3 4 5 6 7 8 9	Topography, Slope & Soil endurance
Access to health care use	9 8 7 6 5 4 3 2 1 2 3 4 5 6 7 8 9	Efficient roads & public transportation (accessibility routes)
Access to commercial use and retails	9 8 7 6 5 4 3 2 1 2 3 4 5 6 7 8 9	Efficient roads & public transportation (accessibility routes)
Access to educational use	9 8 7 6 5 4 3 2 1 2 3 4 5 6 7 8 9	Efficient roads & public transportation (accessibility routes)
Rescue and security facilities, the number of crisis headquarters centers	9 8 7 6 5 4 3 2 1 2 3 4 5 6 7 8 9	Efficient roads & public transportation (accessibility routes)
Pipelines, urban equipment, access to and supply of water sources	9 8 7 6 5 4 3 2 1 2 3 4 5 6 7 8 9	Efficient roads & public transportation (accessibility routes)
Access to open space	9 8 7 6 5 4 3 2 1 2 3 4 5 6 7 8 9	Efficient roads & public transportation (accessibility routes)

Sub-criteria (elements)	Priorities	Sub-criteria (elements)
Access to public space	9 8 7 6 5 4 3 2 1 2 3 4 5 6 7 8 9	Efficient roads & public transportation (accessibility routes)
Access to commercial use and retails	9 8 7 6 5 4 3 2 1 2 3 4 5 6 7 8 9	Access to health care use
Access to educational use	9 8 7 6 5 4 3 2 1 2 3 4 5 6 7 8 9	Access to health care use
Rescue and security facilities, the number of crisis headquarters centers	9 8 7 6 5 4 3 2 1 2 3 4 5 6 7 8 9	Access to health care use
Pipelines, urban equipment, access to and supply of water sources	9 8 7 6 5 4 3 2 1 2 3 4 5 6 7 8 9	Access to health care use
Access to open space	9 8 7 6 5 4 3 2 1 2 3 4 5 6 7 8 9	Access to health care use
Access to public space	9 8 7 6 5 4 3 2 1 2 3 4 5 6 7 8 9	Access to health care use
Access to educational use	9 8 7 6 5 4 3 2 1 2 3 4 5 6 7 8 9	Access to commercial use and retails
Rescue and security facilities, the number of crisis headquarters centers	9 8 7 6 5 4 3 2 1 2 3 4 5 6 7 8 9	Access to commercial use and retails
Pipelines, urban equipment, access to and supply of water sources	9 8 7 6 5 4 3 2 1 2 3 4 5 6 7 8 9	Access to commercial use and retails
Access to open space	9 8 7 6 5 4 3 2 1 2 3 4 5 6 7 8 9	Access to commercial use and retails
Access to public space	9 8 7 6 5 4 3 2 1 2 3 4 5 6 7 8 9	Access to commercial use and retails
Rescue and security facilities, the number of crisis headquarters centers	9 8 7 6 5 4 3 2 1 2 3 4 5 6 7 8 9	Access to educational use
Pipelines, urban equipment, access to and supply of water sources	9 8 7 6 5 4 3 2 1 2 3 4 5 6 7 8 9	Access to educational use
Access to open space	9 8 7 6 5 4 3 2 1 2 3 4 5 6 7 8 9	Access to educational use
Access to public space	9 8 7 6 5 4 3 2 1 2 3 4 5 6 7 8 9	Access to educational use
Pipelines, urban equipment, access to	9 8 7 6 5 4 3 2 1 2 3 4 5 6 7 8 9	Rescue and security facilities (Emergency, Police,

Sub-criteria (elements)	Priorities																	Sub-criteria (elements)
and supply of water sources																		Firefighting), The number of crisis headquarters centers
Access to open space	9	8	7	6	5	4	3	2	1	2	3	4	5	6	7	8	9	Rescue and security facilities, The number of crisis headquarters centers
Access to public space	9	8	7	6	5	4	3	2	1	2	3	4	5	6	7	8	9	Rescue and security, The number of crisis headquarters centers
Access to open space	9	8	7	6	5	4	3	2	1	2	3	4	5	6	7	8	9	Pipelines, urban equipment, access to and supply of water sources
Access to public space	9	8	7	6	5	4	3	2	1	2	3	4	5	6	7	8	9	Pipelines, urban equipment, access to and supply of water sources
Access to public space	9	8	7	6	5	4	3	2	1	2	3	4	5	6	7	8	9	Access to open space

Author details

Mehrnaz Ramzanpour[1] and Rouhollah Rahimi[2]*

1 Faculty of Architecture and Urban Planning, Shahid Rajaee Teacher Training University, Tehran, Iran

2 Faculty of Arts and Architecture, Department of Architecture, University of Mazandaran, Babolsar, Iran

*Address all correspondence to: r.rahimi@umz.ac.ir

References

[1] Rahimi R, Ansari M, Bemanian M, Mahdavinejad M. Relation between sense of belonging to place and participation on promoting social capital in informal settlements (Case study: Tork-Mahalleh neighborhood at the city of Sari). Journal of Sustainable Architecture and Urban Design. 2020;**8** (1):15-29. DOI: 10.22061/jsaud.2020. 4618.1364

[2] Mohammadi C et al. An analysis of the vulnerability of the city of Kermanshah against the risk of earthquakes with an emphasis on. Journal of Environmental Science Studies. 2023;**8**(2):6338-6350

[3] Sharghi A, Nourtaghani A, Ramzanpour M, Gorji RB. Low-income housing location based on affordable criteria using AHP model and GIS technique (case study: Babolsar City). Environment, Development and Sustainability. 2022;**24**(8):10337-10377

[4] Vale LJ, Shamsuddin S, Gray A, Bertumen K. What affordable housing should afford: Housing for resilient cities. City. 2014;**16**(2):21-50

[5] Chen M, Zhang W, Lu D. Examining spatial pattern and location choice of affordable housing in Beijing, China: Developing a workable assessment framework. Urban Studies. 2015;**52**(10): 1846-1863

[6] Simmie J, Martin R. The economic resilience of regions: Towards an evolutionary approach. Cambridge Journal of Regions, Economy and Society. 2010;**3**:27-43

[7] Lawrence RJ. Housing and health: from interdisciplinary principles to transdisciplinary research and practice. Futures. 2004;**36**(4):487-502

[8] White I. Water and the City: Risk, Resilience and Planning for a Sustainable. Future: Routledge; 2013

[9] Mileti D. Disasters by design: A reassessment of natural hazards in the United States. Washington, D.C.: Joseph Henry Press; 1999

[10] Holling CS. Resilience and stability of ecological systems. Annual review of ecology and systematics. 1973;**4**(1):1-23. DOI: 10.1007/s10464-007-9156-6

[11] Bousquet F, Botta A, Alinovi L, Barreteau O, Bossio D, Brown K, et al. Resilience and development: mobilizing for transformation. Ecology and Society. 2016;**21**(3):40-58

[12] Woodcraft S, Bacon N, Caistor-Arendar L, Hackett T. Design for Social Sustainability: A Framework for Creating Thriving New Communities. London, England: Social Life/Young Foundation; 2012

[13] Masten AS. Global perspectives on resilience in children and youth. Child Development. 2014;**85**(1):6-20. DOI: 10.1111/cdev.12205

[14] Cutter SL. The landscape of disaster resilience indicators in the USA. Natural Hazards. 2016;**80**(2):741-758. DOI: 10.1007/s11069-015-1993-2

[15] Dowell SF, Blazes D, Desmond-Hellmann S. Four steps to precision public health. Nature. 2016;**540**(7632): 189-191

[16] Ungar M. Which counts more: Differential impact of the environment or differential susceptibility of the individual? British Journal of Social Work. 2017;**47**(5):1279-1289

[17] Annarelli A, Nonino F. Strategic and operational management of organizational resilience: Current state of research and future directions. Omega. 2016;**62**:1-18. DOI: 10.1016/j.omega.2015.08.004

[18] Walker B, Salt D, Reid W. Resilience Thinking: Sustaining Ecosystems and People in a Changing World. Washington, DC: Island Press; 2006

[19] Hassler U, Kohler N. Resilience in the built environment. Building Research & Information. 2014;**42**(2):119-129. DOI: 10.1080/09613218.2014.873593

[20] Pickett STA, Cadenasso ML, Grove JM. Resilient cities: Meaning, models, and metaphorfor integrating the ecological, socio- economic, and planning realms. Landscape and Urban Planning. 2003;**69**(4):369-384. DOI: 10.1016/j.landurbplan.2003.10.035

[21] Belsky ES, Goodman J, Drew R. Measuring the Nation's Rental Housing Affordability Problems. Cambridge, MA: Joint Centre for Housing Studies, Harvard University; 2005 Available from: http://www.jchs.harvard.edu/research/publications

[22] Rowley S. Housing affordability, housing stress and household wellbeing in Australia. 2012. Available from: melbourneinstitute.unimelb.edu.au

[23] Gabriel M, Jacobs K, Arthurson K, Burke T, Yates J. Conceptualising and Measuring the Housing Affordability Problem, National Research Venture 3: Housing Affordability for Lower Income Australians. Research paper, 1. Australian Housing and Urban Research Institute. 2005. ISBN: 1 902941 76 2 [Report]

[24] Saaty TL. The Analytic Hierarchy Process: Planning, Priority Setting, Resource Allocation. New York and London: McGraw-Hill International Book Co; 1980

[25] Saaty TL. Highlights and critical points in the theory and application of the analytic hierarchy process. European Journal of Operational Research. 1994; **74**:426-447

[26] Delshad M, Tabibian M, Habibi SM. Investigation of the concept of spatial-physical resilience in relation to earthquake, through introducing and prioritizing its most significant criteria by using fuzzy-AHP model (case study: The central texture of the city of Rasht). Armanshahr Architecture & Urban Development. 2021;**14**(36):204-223

[27] Farhadi E, Pourahmad A, Ziari K, Faraji Sabokbar H, Tondelli S. Indicators affecting the urban resilience with a scenario approach in Tehran metropolis. Sustainability. 2022;**14**:12756. DOI: 10.3390/su141912756

[28] Ghasemi M, Gharaee M. Investigation of physical resilience of suburban housing against natural hazards (Case study: Bahonar town of Mashhad). Disaster Prevention and Management Knowledge. 2021;**11**(1):38-54

[29] Lanagarneshin A, Arghan A, Karkehabadi Z. Measurement of Environmental physical indicators of resilience. In: Urban Texture of Tehran (Case study: Tajrish, Jenatabad, Ferdowsi, Tehran) In order to provide a native model for resilient macro-cities in Iran. Geography (Regional Planning). 2019;**9**(2):669-693

[30] Pashapour HA, Pourakrami M. Measuring physical dimensions of urban resilience in the face of the natural disasters (Earthquake) (Case study: Tehran's 12th District). Journal of Studies of Human Settlements Planning. 2018;**12**(4):985-1002

[31] Abdolah B et al. Evaluating the physical resilience in urban districts using ELECTRE technique (case study: region 5, Tehran City). Geography (Regional Planning). 2020;**10**(2–2): 1125-1139

[32] Rezaie E, Salehi E, Lahijanian A, Ehsani A. Determining criteria and measuring resilience status of urban land use in district 22 of Tehran. Physical Geography Quarterly. 2022;**14**(54): 127-146

[33] Desouza KC, Flanery TH. Designing, planning, and managing resilient cities: A conceptual framework. Cities. 2013;**35**: 89-99

[34] Chen Y, Liu T, Ge Y, Xia S, Yuan Y, Li W, et al. Examining social vulnerability to flood of affordable housing communities in Nanjing, China: Building long-term disaster resilience of low-income communities. Sustainable Cities and Society. 2021;**71**:102939

[35] Adger N. Social and ecological resilience: Are they related? Progress in Human Geography. 2000a;**24**(3):347-364

[36] Adger WN, Kelly PM, Winkels A, Huy LQ, Locke C. Migration, remittances, livelihood trajectories, and social resilience. Ambio: A Journal of the Human Environment. 2002;**31**(4): 358-366

[37] Assarkhaniki Z, Rajabifard A, Sabri S. The conceptualisation of resilience dimensions and comprehensive quantification of the associated indicators: A systematic approach. International Journal of Disaster Risk Reduction. 2020;**51**:101840

[38] Lyon C. Place systems and social resilience: A framework for understanding place in social adaptation, resilience, and transformation. Society & National Resources. 2014;**27**(10): 1009-1023

[39] Orencio PM, Fujii M. A localized disaster-resilience index to assess coastal communities based on an analytic hierarchy process (AHP). International Journal of Disaster Risk Reduction. 2013; **3**:62-75

[40] Pokhrel S. Green space suitability evaluation for urban resilience: an analysis of Kathmandu Metropolitan city, Nepal. Environmental Research Communications. 2019;**1**(10):105003

[41] Zhang M, Yang Y, Li H, van Dijk MP. Measuring urban resilience to climate change in three chinese cities. Sustainability. 2020;**12**(22):9735

[42] Adger N. Migration, remittances, livelihood trajectories and social resilience. Ambio: A Journal of the Human Environment. 2000b;**31**(4): 358-366

[43] Ebrahimnia V, Shahbazi T, Rahimi R. The analysis of spatial environment influencing social capital in neighborhoods of Sari City. Geography and Urban Space Development. 2022;**9**(3):121-140. DOI: 10.22067/jgusd.2022.67168.0

[44] Norris FH, Stevens SP, Pfefferbaum B, Wyche KF, Pfefferbaum RL. Community resilience as a metaphor, theory, set of capacities, and strategy for disaster readiness. American Journal of Community Psychology. 2008;**41**:127-150

[45] Partovi P, Behzadfar M, Shirani Z. Urban design and social resiliency case study: Jolfa neighborhood in Isfahan City. Journal of Architecture and Urban Planning. 2016;**9**(17):99-116

[46] Abdullah et al. Evaluated the physical resilience in Tehran (Iran) with

the indicators of population and building density, roads, worn-out texture, green and open spaces, health centers, crisis headquarters centers, and natural bed characteristics. 2020

[47] Lam K, Zhao X. An application of quality function deployment to improve the quality of teaching. International Journal of Quality and Reliability Management. 1998;**15**(4):389-413

[48] Tavares RM, Tavares JL, Parry-Jones SL. The use of a mathematical multi-criteria decision-making model for selecting the fire origin room. Building and Environment. 2008;**43**(12): 2090-2100

[49] Cheng EW, Li H, Ho DC. Analytic hierarchy process (AHP): A defective tool when used improperly. Measuring Business Excellence. 2002;**6**(4):33-37

[50] Akadiri PO, Olomolaiye PO, Chinyio EA. Multi-criteria evaluation model for the selection of sustainable materials for building projects. Automation in Construction. 2013;**30**:113-125

[51] Lam KC, Lam MCK, Wang D. MBNQA-oriented self-assessment quality management system for contractors: fuzzy AHP approach. Construction Management and Economics. 2008;**26**(5):447-461

[52] Li J, Zou PXW. Fuzzy AHP-based risk assessment methodology for PPP projects. Journal of Construction Engineering and Management. 2011;**137** (12):1205-1209

[53] Ali HH, Al Nsairat SF. Developing a green building assessment tool for developing countries–Case of Jordan. Building and environment. 2009;**44**(5): 1053-1064

[54] Oluleye IB, Ogunleye MB, Oyetunji AK. Evaluation of the critical success factors for sustainable housing delivery: Analytic hierarchy process approach. Journal of Engineering, Design and Technology. 2021;**19**(5):1044-1062

[55] Yang J, Lee H. An AHP Decision Model for Facility Location Selection. Facilities. 1997;**15**:241-254

[56] Saaty TL. Decision making with the analytic hierarchy process. International journal of services sciences. 2008;**1**(1): 83-98

[57] Alonso JA, Lamata MT. Consistency in the analytic hierarchy process: A new approach. International Journal of Uncertainty, Fuzziness and Knowledge-Based Systems. 2006;**14**(04):445-459

[58] Soltani-Sobh A, Heaslip K, Scarlatos P, Kaisar E. Reliability based pre-positioning of recovery centers for resilient transportation infrastructure. International Journal of Disaster Risk Reduction. 2016;**19**:324-333. DOI: 10.1016/j.ijdrr.2016.09.004

[59] Zhang W, Wang N. Resilience-based risk mitigation for road networks. Structural Safety. 2016;**62**:57-65. DOI: 10.1016/j.strusafe.2016.06.003

[60] Zhang X, Miller-Hooks E, Denny K. Assessing the role of network topology in transportation network. Journal of Transport Geography. 2015;**46**:35-45. DOI: 10.1016/j.jtrangeo.2015.05.006

[61] Aminzadeh B, Adeli Z. Evaluation of the urban fabric vulnerability in the earthquake crisis (case study Qazvins regions). Hoviatshahr. 2015;**8**(20):5-18

[62] Tabibian M, Mozafari N. Assessment of vulnerability of residential areas to earthquake disasters and it's planning guidelines (Case study: District number 6, Tehran municipality). Motaleate Shahri. 2018;**7**(27):93-112

Chapter 5

The Problem of the Adequacy of the Analytic Hierarchy Process and Its Solution

Vasili Romanchak

Abstract

The Analytic Hierarchy Process (AHP) is a popular method for solving multi-criteria problems. Many researchers emphasize the simplicity and naturalness of the AHP procedure for evaluating alternatives. However, many scholars believe that the AHP is flawed and therefore cannot be applied in practice. Such a scattering of opinions requires an explanation. The method of AHP uses pairwise comparisons of alternatives. And it is based on the assumption that the alternatives can be measured on the scale of relations. Fechner's psychophysical law is used as justification for the existence of method of measurement with ratio scales. But there are not one but two psychophysical laws. The existence of two psychophysical laws is a problem of psychophysics. This problem has been solved quite recently. It was shown that the basic psychophysical laws are equivalent. In order to solve this problem, an adjustment of Stevens' direct measurement model was required. It is suggested that a direct measurement model be used to overcome the shortcomings of the Analytic Hierarchy Process. In this case, the fundamental AHP measurement scale can be used. The adjusted AHP method contains a direct measurement model and a built-in mechanism for checking the adequacy of measurement results. An example of direct measurement of alternatives is analyzed.

Keywords: Fechner's law, Stevens' law, analytic hierarchy process (AHP), rating, theory of measurement

1. Introduction

The Analytic Hierarchy Process (AHP) evaluates alternatives using pairwise comparisons based on expert judgments [1, 2]. The AHP methodology involves the measurement of alternatives and the transformation of measurement results [3]. Using AHP methods, it is possible to construct a mathematical model of decision-making [4, 5]. Since the AHP method does not have a strict mathematical justification, there are various modifications to the method [6–8]. The Multiattribute Utility Theory (MAUT) [9–11] and specially created methods [12, 13] can be used instead of the AHP method.

In the monograph [14], numerous examples of AHP applications were considered. The authors of the monograph concluded that, despite its popularity, AHP is incapable of solving complex problems. The popularity of AHP is because the AHP method gives the researcher the feeling that he or she is actually solving a complex problem based on his or her preferences. Therefore, many AHP users consider that this method can be applied to any scenario. The authors of the monograph believe that, even when solving trivial problems, AHP is based on questionable procedures. There are references to the works of more than 100 scientists who support this view.

J. Barzilai is the author of a New Theory of measurement [15]. J. Barzilai set the conditions for using the mathematical operations of linear algebra and calculus. The failure to meet the conditions for the application of arithmetic operations in the mathematical foundations of measurement theory, utility theory, and decision theory caused fundamental errors [16]. J. Barzilai believes that T. Saaty did not define what was meant by the terms "importance of criteria" or "relative importance" of criteria. Moreover, criterion importance coefficients cannot be interpreted as a measure of the relative importance of criteria. "In fact, the AHP is plagued by many flaws, and these flaws are fundamental" [17]. There are many examples of the method not working correctly [18]. However, the AHP method is still popular. This can be explained by the fact that the AHP contains a direct measurement model, which is absent in axiomatic theories of measurement. In addition, the AHP measurement model considers the psychological features of the person.

The purpose of this paper is to propose a modification of the AHP that retains the AHP measurement scale. In this case, the decision-making model is free from fundamental errors. The possibility of such modification is explained by the fact that the justification of the AHP method derives from Fechner's psychophysical law. It should be emphasized that there are not one but two psychophysical laws in psychophysics: Fechner's law and Stevens' law. The existence of two psychophysical laws was until recently considered a problem [19–21]. A solution to this problem was obtained in [22–24] through the application of the Stevens measurement model. In this paper, this measurement model is used to adjust the AHP method. For a specialist, the AHP correction comes down to small changes in the calculation scheme.

At the beginning of the article, the AHP measurement model, which is the basis of the AHP method, is briefly described. An analysis of the measurement model indicates that the model needs to be adjusted. The Stevens measurement model is then discussed, and the multifactor model is justified. The application of the new measurement model is then described. As an illustration, the paper considers an example of a proper evaluation of alternatives using the AHP scale. Finally, the advantages of the proposed approach are presented.

2. A critique of the analytic hierarchy process as a method of measurement with ratio scales

The measurement model was defined by T. Saaty using the results of Fechner's works [1]. Fechner (1860) investigated the reactions that arise when paired comparisons of stimuli are made. For example, in an experiment, participants are offered two objects of a certain weight. The weight of one object is changed until the participants notice a difference. Fechner called such a difference "just noticeable."

Fechner believed that for a sequence of "just noticeable differences" in stimuli, the relation $v_i = v_{i-1}\alpha$ is satisfied, where v_i is the value of the stimulus and v_0 is the numerical value of the first stimulus, and α is a constant. Then the equality $v_i = v_0\alpha^i$, and $v_j = v_0\alpha^j$ are fulfilled, and equality is obtained

$$\ln(\alpha)(u_i - u_j) = \ln(v_i/v_j), \tag{1}$$

where $(u_i - u_j) = (i - j)$. Equality (1) demonstrates that the ratio of values of the physical quantity v_i/v_j can be found by subjectively measuring the difference in values of $(u_i - u_j)$.

If the result of the measurement is the difference of values, the values are determined to be the additive constant. Assume $u_1 - u_0 = 1, u_2 - u_1 = 2$. Then $u_0 = C, u_1 = 1 + C, u_2 = 2 + C$, where C is a constant. Only one arithmetic operation is defined for the values u_0, u_1, and u_2 : the difference of values. The ratio of these values is undefined. T. Saaty believed that if the constant C is set to zero, the values are uniquely defined: $u_0 = 0$, $u_1 = 1$, and $u_2 = 2$, and the division operation is defined. Indeed, in this case, the division operation is formally defined. But this operation does not make empirical sense. For example, there is no reason to suppose that the value of u_2 is twice the value of u_1. Thus, the justification of the AHP method looks unconvincing.

A similar modeling error is observed when forming a matrix of pairwise comparisons using the AHP method. A matrix of pairwise comparisons is the basic element of the AHP [1]. Consider a set of alternatives $A_1, A_2, \dots, A_n$. Weights of alternatives $w_j = w(A_j)$ are found as a result of pairwise comparison of alternatives. The results of pairwise comparisons are represented as a square matrix $V_{ij} = [v_{ij}]$, $i, j = 1, 2, \dots, n$, where $v_{ij} = w_i/w_j$ is an estimate of the relative importance of alternative A_i compared to alternative A_j and $v_{ij} = 1/v_{ji}$. The weights of the alternatives are unknown and are determined by the results of pairwise comparisons. In the hierarchy analysis method, the results of pairwise comparisons are chosen from an integer "fundamental" scale of {1, 2, ..., 9}.

If a matrix of pairwise comparisons is given, the weights of the alternatives are not uniquely defined. For example, let $v_{21} = 2$, $v_{32} = 2$, and $v_{31} = 4$. Then the system of equations has the form: $w_2/w_1 = 2, w_3/w_2 = 2$, $w_3/w_1 = 4$. This system of equations can be solved as follows, where C is an unknown constant: $\ln(w_1) = C$, $\ln(w_2) = C + \ln(2)$, $\ln(w_3) = C + \ln(4)$.Hence, the weights are defined on log-intervals scale. If the constant C is fixed, then a division and addition operation is formally defined for the values of the quantity. For example, if $C = 0$, then $w_1 + w_2 = 3$, but this operation has no empirical justification. However, the sum of weights is used in the AHP. Therefore, the AHP uses arithmetic operations without proper justification. Since arithmetic operations do not correspond to empirical operations, it violates the principle of reflection [15].

The aim of the work is to modify the AHP method based on the correct model for measuring alternatives. To this end, the paper considers a mathematical model of the empirical system (J. Bazilai [15]) and a model of direct measurement.

3. The correction of Stevens' scales of measurement (direct measurement model)

Let the empirical system be a straight line with a set of points and a set of vectors [15]. For any ordered pair of points A_1 and A_2 there is a unique vector (A_1, A_2). Points

Figure 1.
Empirical system. Measurement objects $A_1, A_2, \dots, A_n$.

on a straight line are measurement objects (**Figure 1**). A vector is the empirical result of a measurement that characterizes the difference in the position of two points on a straight line.

Thus, the model of the empirical system is a one-dimensional affine space. J. Barzilai calls this space homogeneous because, in this case, it is possible to compare vectors with each other [15]. For example, according to **Figure 1**, the vector (A_1, A_2), is two times smaller than the vector (A_1, A_3).

Following A. Friedman (1922), let us axiomatically define an "exceptional group of objects" admitting a special estimation [25]. Let us assume that objects $A_1, A_2, \dots, A_n$ are arranged in increasing order of the measured value and the value of objects changes uniformly. This means that the vectors (A_1, A_2), (A_2, A_3), …, (A_{n-1}, A_n) are equal to each other. The founder of the representative theory of measurement, S. S. Stevens, used a similar model for direct measurement of value. Let the value of the quantity u_i corresponds to object A_i. Consider that successive value differences are equal to one another:

$$u_2 - u_1 = u_3 - u_2 = \dots = u_n - u_{n-1}. \tag{2}$$

A.A. Friedman called such special estimation "measurement." Then

$$u_i - u_j = \lambda_1(i - j), \tag{3}$$

when λ_1 is an unknown constant, $\lambda_1 > 0$. This implies that the values of the measured quantity are defined with an accuracy of a linear transformation, that is, in the interval scale. Assign the values of the quantity v_i to each object A_i, and assume that the successive ratios are equal

$$v_2/v_1 = v_3/v_2 = \dots = v_n/v_{n-1}. \tag{4}$$

The equality

$$\ln(v_i/v_j) = \lambda_2(i - j), \tag{5}$$

is then satisfied, where λ_2 is an unknown constant, $\lambda_2 > 0$. The values of the logarithms of the measured quantity are determined with an accuracy of a linear transformation, i.e., on the log-interval scale [26].

Thus, two ways of measuring the value are obtained. In the first case, the result of the measurement is the difference $(u_i - u_j)$, and in the second case, the ratio of the values (v_i/v_j). The values of a quantity are measured on a scale of intervals (3) or log intervals (5) [26, 27].

Similar models were used by C. S. Stevens to classify measurement scales. The choice of four measurement scales was made by S.S. Stevens back in 1946 [26]. S.S. Stevens later added a fifth scale to them, the scale of logarithmic intervals, but it was later recognized as useless [27]. At first glance, Stevens' concept of measurement looks convincing, and only the presence of an "extra" fifth scale violates the logic of the presentation. According to S.S. Stevens, the scale of logarithmic intervals is mathematically interesting but, like many mathematical models, empirically useless. Let us use an example to demonstrate why such a claim is controversial. To do this, let us measure a non-additive quantity using the Stevens model.

Density is an example of a non-additive value. For example, if the density values of two samples are 3 kg/m^3 and 2 kg/m^3, it is not clear what the sum of these values would mean. But the division operation is defined for density; specifically, 3 kg/m^3 is 1.5 times greater than 2 kg/m^3. Let the densities of five samples A_1, A_2, A_3, A_4, A_5 change uniformly, and for clarity, let the ratio of the densities of two consecutive samples be two. Density values can be compared in two ways. To calculate the difference in values, use formula (3): $u_i - u_j = (i - j)$, where u_i is the density value. Using formula (5), the ratio $v_i/v_j = 2^{i-j}$, is obtained, where$i, j = 1, 2, \ldots, 5$.

The density values v_i are determined by the accuracy of the multiplier; valuesu_i are determined up to the additive constant. In this particular case, the values are given in **Table 1**. The values have a natural interpretation. For example, the density of the third sample is four times greater than the first, or two orders of magnitude greater than the first.

The example confirms that for arbitrary objects $A_1, A_2, \ldots, A_n$, the value of which *changes uniformly*, it is reasonable to consider two measurement scales: the scale of intervals and the scale of log intervals.

The ratio scale is the highest level of measurement in Stevens' classification of levels of measurement (nominal, ordinal, interval, and ratio) [26–28]. The ratio scale is invariant over transformations in which the numerals on the scale are multiplied by a constant. From the analyzed example, it follows that the result of direct measurement is the scale of intervals or the scale of log intervals (**Table 1**). Therefore, the ratio scale is not a scale of direct measurement. But S.S. Stevens believed that the log-interval scale was useless [27] and carried out direct measurements on the ratio scale. This point of view was considered correct for a long time and was the cause of numerous errors, including those in the AHP method.

4. Adjusted model of direct measurement

From equalities (3) and (5), it follows that the values of the quantity on the scale of intervals and log intervals are related by the formula

$$\left(u_i - u_j\right) = \lambda \ln\left(v_i/v_j\right). \tag{6}$$

where $i, j = 1, 2, \ldots, n$; u_i, v_i are values of the quantity and $\lambda = \lambda_1/\lambda_2$. Equality (6) holds true for the values u_i and v_i in **Table 1**, if $\lambda = 1/\ln(2)$.

u_i	1	2	3	4	5
v_i	2	2^2	2^3	2^4	2^5

u_i are "values" in the interval scale,
v_i are values in the log-interval scale.

Table 1.
Density values.

Equality (6) means that the mapping $u = \ln(v)$ preserves the operation of measurement: the ratio of values maps to the difference of values. In addition, there is a one-to-one correspondence between the values of u_i and v_i. The mapping $u = \ln(v)$ is an isomorphism of two algebraic structures: the set of all positive numbers with the operation of division, and the set of all real numbers with the operation of subtraction. In algebra, isomorphic structures are not distinguished; they are equivalent [29].

Let in the process of measurement, each pair of objects is assigned a difference $u_i - u_j$ or ratio v_i/v_j of values. In order to consider two methods of measuring simultaneously, let us denote the left and right parts of equality (6) by the symbol R_{ij} and define two mappings

$$R_{ij} = \lambda_1 (u_i - u_j), \tag{7}$$

$$R_{ij} = \lambda_2 \ln(v_i/v_j), \tag{8}$$

where R_{ij} is the *values of the rating*; $i, j = 1, 2, \dots, n$; u_i and v_i are the values of quantity, and λ_1, λ_2 are positive constants. For objects whose value changes uniformly, the rating is defined by the formula $R_{ij} = \lambda(i - j), \lambda > 0$. This definition of rating is called "classical" [22]. The rating does not depend on the choice of measurement method (7) or (8).

It is possible to check directly that the rating values R_{ij} satisfy the compatibility condition of the form

$$R_{ij} = R_{ik} + R_{kj}, \tag{9}$$

In [22–24], the compatibility condition (9) is axiomatically defined, and the theoretical model of direct measurement is formulated.

The values of a quantity are obtained on a scale of intervals if they are the solution of the system of Eq. (7), and on a scale of log intervals if they are the solution of the system of Eq. (8). For example, if the respondent believes that the criterion A_3 exceeds the criterion A_1 by six conditional units and the criterion A_2 exceeds criterion A_1 by three conditional units, then the equations $u_3 - u_1 = 6$ and $u_2 - u_1 = 3$ are true and values of u_1, u_2, and u_3 are determined on an interval scale. If the respondent believes that criterion A_3 is four times more important than criterion A_1 and criterion A_2 is two times more important than criterion A_1, then the system of equations looks like this: $v_3/v_1 = 4$ и $v_2/v_1 = 2$. In this case, the values are determined using a logarithmic interval scale. The values on the interval scale and the log-interval scale must be related by the equivalence condition (6).

Various measurement models have been widely used for a long time, but the direct measurement model includes isomorphism (equivalence condition (6)) of scales. From the equivalence condition, follow Fechner's law in the form of paired comparisons [22], Stevens' law in the form of paired comparisons [22], and Rush's model [30, 31].

Stevens' experimental law (1947) was proposed instead of Fechner's experimental law (1848). There is now a paradoxical contradiction in psychophysics between Fechner's and Stevens' laws, in that the two basic laws contradict each other. The harmonization of these two laws has been the subject of much discussion, but a solution that would satisfy all involved has never been found [19]. The fact that the direct measurement model under consideration solves the complex problem of psychophysics confirms its theoretical and practical importance. Using the direct measurement model, it is possible to introduce the notion of independent variables.

5. Independence of variables

Let A and B be two sets of real numbers and let D be the Cartesian product of sets A and $B, D = A \times B$. Each pair of values (a, b) from D corresponds to the value of the function $u = u(a, b)$. Let $M_0(a_0, b_0)$ is point from the set D. The total increment of the function $u = u(a, b)$ at the point $M_0(a_0, b_0)$ is the difference

$$\Delta u = u(a, b) - u(a_0, b_0). \tag{10}$$

For a fixed point, $M_0(a_0, b_0)$ the total increment z is a function of the variables (a, b), $\Delta u = \Delta u(a, b)$. The partial increment of the function $u = u(a, b)$ by the variable is the difference

$$\Delta_a u(a, b) = u(a, b) - u(a_0, b), \tag{11}$$

$$\Delta_b u(a, b) = u(a, b) - u(a, b_0). \tag{12}$$

***Definition.* Two variables are called independent if the partial increment of each of them does not depend on what value the other variable has taken:**

$$\Delta_a u(a, b) = \Delta_a u(a), \tag{13}$$

$$\Delta_b u(a, b) = \Delta_b u(b). \tag{14}$$

The problem of finding the total increment of a function based on partial increments has no solution in the general case. In the special case where the variables are independent, the problem is solved quite simply. In this case, the total increment of a function is equal to the sum of the partial increments

$$\Delta u(a, b) = \Delta u_a(a) + \Delta u_b(b). \tag{15}$$

Let $\Delta u_a(a) = k_1 x$ and $\Delta u_b(b) = k_2 y$then equality (15) can be written as

$$\Delta u(x, y) = k_1 x + k_2 y, \tag{16}$$

where $\Delta u(x, y) = u(x, y) - u(x_0, y_0); k_1$ and k_2 are scale constants. Similarly, the definition of independent variables can be formulated on the basis of a direct measurement of the ratio of values. In this case equality

$$\delta u(x, y) = k_1 x + k_2 y \tag{17}$$

is satisfied, where $\delta u(x, y) = \ln(v(x, y)/v(x_0, y_0))$. Modified AHP method uses an additive representation of the form (16) or (17).

6. Modified analytical hierarchy process (example)

Let three main factors influence the favorable sociopolitical development of the state: x, economy; y, ecology; z, security. Each factor has a certain number of discrete levels. (The example is partially taken from the monograph [1]).

The project is a point (x, y, z) in three-dimensional space, the coordinates of which correspond to the level of development of the economy, ecology, and security. Let the value of the function $U(x, y, z)$ be a numerical characteristic of the state's development. Factors $x, y,$ and z can be considered mutually independent. By analogy with the representation (16), let us choose an additive model to assess the development of the state

$$U - U_0 = k_1 x + k_2 y + k_3 z, \tag{18}$$

where $k_1, k_2,$ and k_3 are coefficients of influence of factors; $x, y,$ and z are variables (factors). Let us denote the lower and upper levels of each factor as 0 and 1, $0 \leq x \leq 1, 0 \leq y \leq 1,$ and $0 \leq z \leq 1$.

6.1 Influence of factors

The projects that need to be considered here are $A_0(0, 0, 0)$, $A_1(1, 0, 0)$, $A_2(0, 1, 0)$, and $A_3(0, 0, 1)$. Let us introduce the notation $U_i = U(A_i)$, $i = 0, 1, 2, 3$; U_i is a numerical characteristic of state development. The value of U_0 corresponds to the current level of the State, and the value of U_1 corresponds to the project with the maximum development of only the economy. The decision maker performs the pairwise comparison of projects. The results of the pairwise comparisons are recorded in the form of a matrix. The matrix of pairwise comparisons according to the AHP method [1] is presented in **Table 2**.

The first row of **Table 2** contains the number 5. This means that the equality $U_1/U_2 = 5$ is satisfied, i.e., the influence of the first factor is five times greater than the influence of the second factor. In the corrected version of the AHP method, it is necessary to replace the ratios in this matrix with differences. For example, if the number 5 appears in the pairwise comparison matrix in **Table 2**, replace it with the difference (5–1). Similarly, if the ratio is 1/ 5, replace it with a difference of 1 and 5. As a result, $U_1 - U_2 = 4$ and $U_2 - U_1 = -4$ are obtained. Equations $U_3 - U_2 = 1$ and $U_1 - U_3 = 2$ are obtained similarly.

To find the influence coefficients $k_1, k_2,$ and k_3 in the formula (18), the data in **Table 2** is not enough. Therefore, the expert must additionally evaluate the influence of each factor using the fundamental scale of the AHP [1]. Let the equalities $U_1 - U_0 = 6, U_2 - U_0 = 2, U_3 - U_0 = 2$ be obtained with the help of an expert. This means that the expert estimates the influence of the first factor to be four units and the influence of the second and third factors to be two units. The result is an extended matrix of pairwise comparisons (**Table 3**).

The result of the pairwise comparison of projects A_i andA_j is the difference in values (rating): $U_{ij} = U_i - U_j, U_{ij} = -U_{ji}$.

U_i/U_j	U_1	U_2	U_2
U_1	1	5	3
U_2	1/5	1	1/2
U_3	1/3	2	1

Table 2.
Matrix of pairwise comparisons (T. Saaty).

U_{ij}	U_0	U_1	U_2	U_3
U_0	0	−6	−2	−2
U_1	6	0	4	2
U_2	2	−4	0	−1
U_3	2	−2	1	0

Table 3.
Extended matrix of pairwise comparisons.

$U_i - U_0$	U^0	U^1	U^2	U^3	$\overline{U}$
U_0	0	0	0	0	0.00
U_1	6	6	6	4	5.50
U_2	2	2	2	1	1.75
U_3	2	4	3	2	2.75

$\overline{U}$ is Average Rating.

Table 4.
Rating of factors.

Columns U^k of **Table 4** are obtained from columns U_k of **Table 3** by subtracting the first element from each element of the column by the formula $U_i^k = U_{ik} - U_{0k}$, $i, k = 0, 1, 2, 3$.

If the rating estimates U_{ij} are found exactly, then all columns U^0, U^1, U^2, and U^3 in **Table 4** must match. Equality in some sense of the columns can be used as a criterion for the adequacy of the measurement results. The vector $\overline{U}$ are found as the arithmetic mean of the columns U^0, U^1, U^2, and U^3 in **Table 4**. The correlation coefficient $\rho_k = \rho(U^k, \overline{U})$, is denoted by ρ_k. In this case, correlation coefficients $\rho_0 = 0.979, \rho_1 = 0.982, \rho_2 = 1.000, \rho_3 = 0.997$are significant at the 0.05 significance level. Therefore, the hypothesis that the mathematical model is adequate for the measurement results is accepted.

6.2 Influence of factor levels

Let $x_0, x_1,$ and x_2 be the levels of the first factor x; y_0, y_1 and y_2 be the levels of the second factory; and, z_0 and z_1 be the levels of the third factor$z$. For example, level x_0, corresponds to the current state of the economy. Let the result of pairwise comparisons be the difference in values of factor levels $x_{ij} = x_i - x_j$. The matrix of pairwise comparisons of levels of the economy (**Table 5**) is made with the fundamental scale AHP.

The number 4 in the first column of the data in **Table 5** means that the level of x_2 is four units higher than the current level of x_0.

Columns $x^0 x^1$, and x^2 of **Table 6** are obtained from columns $x_0, x_1,$ and x_2 of **Table 5** by subtracting the first element from each column element using the formula $x_i^k = x_{ik} - x_{0k}$.

x_{ij}	x_0	x_1	x_2
x_0	0	−1	−3
x_1	1	0	−2
x_2	4	2	0

Table 5.
Pairwise comparisons of Economic Levels.

x	x^0	x^1	x^2	$\overline{x}$	$\hat{x}$
x_0	0	0	0	0.00	0.00
x_1	1	1	1	1.00	0.30
x_2	4	3	3	3.33	1.00

$\overline{x}$ is Average Rating
$\hat{x}$ is Normalized Rating.

Table 6.
Rating of Economic Levels.

y_{ij}	y_0	y_1	y_2
y_0	0	−4	−8
y_1	4	0	−4
y_2	8	4	0

Table 7.
Pairwise comparison of Ecological Levels.

The correlation coefficient is denoted by $\rho_k = \rho(\overline{x}, x^k)$. In this case, correlation coefficients $\rho_0 = 0.999, \rho_1 = 0.999$, and $\rho_2 = 0.999$, are significant at the 0.05 significance level. Therefore, there is reason to believe that the values of the levels correspond to the results of the measurement. Similarly, levels of ecological development are compared to the current level (**Table 7**).

Columns $y^0 y^1$, and y^2 of **Table 8** are obtained from columns y_0, y_1, and y_2 of **Table 7** using the formula $y_i^k = y_{ik} - y_{0k}$.

In this case, all three vectors y^0, y^1, and y^2 coincide (**Table 8**). Therefore, the hypothesis that a normalized rating vector is adequate for the measurement results is accepted. The safety factor z has only two levels. So, let $z_0 = 0$ and $z_1 = 1$.

y	y^0	y^1	y^2	$\overline{y}$	$\hat{y}$
y_0	0	0	0	0	0,0
y_1	4	4	4	4	0,5
y_2	8	8	8	8	1,0

$\overline{y}$ is Average Rating
$\hat{y}$ is Normalized Rating.

Table 8.
Rating of Ecological Levels.

6.3 Project assessment

The linear model (18) allows us to evaluate various social development projects. Let A_m be the project; m be the project number; $m = 0, 1, \dots, 7$; $\left(x_i, y_j, z_k\right)$ be the levels of factors in accordance with **Table 9**. The values of the function $U = U(A_m)$ are estimates of the impact of the projects on the state. The function $U = U(A_m)$ is defined as the linear transformation. To calculate the values of the function, it is convenient to assume that $U(A_0) = 0$ and $U(A_7) = 8$. The rating of the projects (**Table 8**) has been calculated by formula (18).

The outcome of measuring a project's impact on society is its project rating. Rating values, differences, and ratios of rating values have an empirical meaning. In contrast to the AHP method, the rating has a quantitative structure. For example, the task is to choose the best project among A_4,A_5, or A_6. From the analysis of **Table 9**, it follows that the A_6 project should be considered first. And project A_6 has a greater influence than project A_5, by 43%.

7. Conclusions and future research directions

The theoretical foundations of the AHP are currently being criticized, in particular, the correctness of the mathematical model. This paper proposes an adjustment to the mathematical model of the AHP using the Stevens direct measurement model. This allows the AHP method's measurement scale to be used to evaluate alternatives. Statistical criteria can be applied to check the adequacy of the measurement results. The corrected algorithm is no more complicated than the AHP method.

The AHP method uses a variant of the pairwise comparison method. At present, the theoretical foundations of the AHP are being criticized, and in particular, the possibility of measuring preferences on the ratio scale is being questioned. Moreover, the criticism of the AHP method looks quite reasonable. For example, the method of pairwise comparisons has long been used by psychologists, but the values are found to be on an interval scale. The paper shows why the method of pairwise comparisons cannot be used to find values on the ratio scale.

The direct measurement model of psychologist S. Stevens was proposed to measure preferences. The article considers two methods of measurement. The first way is

A_m	x_i	y_j	z_k	$U(A_m)$
A_0	0.00	0.00	0.00	0.00
A_1	1.00	0.00	0.00	4.40
A_2	0.00	1.00	0.00	1.40
A_3	0.00	0.00	1.00	2.20
A_4	0.30	1.00	1.00	4.92
A_5	1.00	0.50	0.00	5.10
A_6	1.00	0.50	1.00	7.30
A_7	1.00	1.00	1.00	8.00

Table 9.
Rating of projects.

to find the difference between values, and the second way is to find the ratio of values. The algorithm for processing measurement results does not change significantly in this case. In this paper, only the first method, which uses the AHP scale, was considered in detail. The values of the quantity in the interval scale were discovered by the method of pairwise comparisons. This approach has certain advantages. In this case, you can use the measurement scale of the AHP method.

The method of pairwise comparisons considered in this paper is fully compatible with the modern theory of measurement by J. Barzilai. Moreover, standard statistical criteria can be used to check the adequacy of the model. As a result, many of the comments made by opponents of the AHP method have been removed. Measurement results obtained by the method of pairwise comparisons can be used to find coefficients of the multiple linear regression model.

It is relevant to continue further research on the corrected AHP method. Future research could include the use of two measurement methods as well as a test of the regression equation's adequacy with measurement results.

Author details

Vasili Romanchak
Belarusian National Technical University, Minsk, Belarus

*Address all correspondence to: romanchak@bntu.by

References

[1] Saaty TL. The Analytic Hierarchy Process. 3rd ed. New York: McGraw-Hill; 1980

[2] Saaty TL. Fundamentals of the analytic hierarchy process. In: The Analytic Hierarchy Process in Natural Resource and Environmental Decision Making. Pittsburgh: RWS Publications; 2000. pp. 15-35

[3] Vaidya OS, Sushil K. Analytic hierarchy process: An overview of applications. European Journal of Operational Research. 2006;**169**(1):1-29

[4] Saaty TL. Decision making with the analytic hierarchy process. International Journal of Services Sciences. 2008;**1**(1): 83-98

[5] Khan AU, Ali Y. Analytical hierarchy process (AHP) and analytic network process methods and their applications: A twenty year review from 2000-2019. International Journal of the Analytic Hierarchy Process. 2020;**12**(3):369-459. DOI: doi.org/10.13033/ijahp.v12i3.822

[6] Leal JE. AHP-express: A simplified version of the analytical hierarchy process method. MethodsX. 2020;7: 100748

[7] Millet I, Saaty TL. On the relativity of relative measures–accommodating both rank preservation and rank reversals in the AHP. European Journal of Operational Research. 2000;**121**(1):205-212

[8] Tung SL, Tang SL. A comparison of the Saaty's AHP and modified AHP for right and left eigenvector inconsistency. European Journal of Operational Research. 1998;**106**(1):123-128

[9] Dyer JS. Multiattribute Utility Theory (MAUT). Multiple Criteria Decision Analysis. New York: Springer; 2016. p. 285

[10] Von Winterfeldt D, Fischer GW. Multi-attribute utility theory: models and assessment procedures. Utility, Probability, and Human Decision Making: Selected. In: Proceedings of an Interdisciplinary Research Conference. Rome: Springer Netherlands; 1975. pp. 47-85

[11] Figueira J et al. MAUT-multiattribute utility theory. Multiple Criteria Decision Analysis: State of the Art Surveys. 2005;**2005**:265-292

[12] De Felice F, Antonella P. Absolute measurement with analytic hierarchy process: A case study for Italian racecourse. International Journal of Applied Decision. 2013;**6**(3):209-227

[13] Zanakis SH et al. Multi-attribute decision making: A simulation comparison of select methods. European Journal of Operational Research. 1998; **107**(3):507-529

[14] Munier N et al. Uses and Limitations of the AHP Method. Berlin: Springer; 2021

[15] Barzilai J. On the foundations of measurement. In: 2001 IEEE International Conference on Systems. Man and Cybernetics. Tucson, AZ: e-Systems and e-Man for Cybernetics in Cyberspace; 2001. pp. 401-406

[16] Barzilai J. Preference function modelling: The mathematical foundations of decision theory. Trends in Multiple Criteria Decision Analysis. 2010;**2010**:57-86

[17] Barzilai J. Notes on the analytic hierarchy process. In: Proceedings of the NSF Design and Manufacturing Research Conference. Tampa, USA; 2001. pp. 1-6

[18] Podinovski VV, Podinovskaya OV. On the theoretical incorrectness of the analytic hierarchy process. Problemy upravleniya. 2011;**1**:8-13

[19] Lubashevsky I. Psychophysical laws as reflection of mental space properties. Physics of Life Reviews. 2019;**31**:276-303

[20] Krueger LE. Reconciling Fechner and Stevens: Toward a unified psychophysical law. Behavioral and Brain Sciences. 1989;**12**(2):251-267

[21] Dehaene S. The neural basis of the Weber–Fechner law: a logarithmic mental number line. Trends in Cognitive Sciences. 2003;**7**(4):145-147

[22] Romancak VM. Measurement of non-physical quantity. System Analysis and Applied Informatics. 2017;**4**:39-44

[23] Romanchak VM. On the equivalence of Fechner's and Stevens' laws. Journal of the Belarusian State University. Philosophy and Psychology. 2019;**3**: 97-102

[24] Romanchak VM, Serenkov PS. Non-additive quantity measurement model. Приборы и методы измерений. 2022; **13**(3):208-215

[25] Friedmann AA. The World as Space and Time. 2nd ed. Moscow: Nauka; 1965. p. 112

[26] Stevens SS. On the theory of scales of measurement. Science. 1946;**103**(2684): 677-680

[27] Stevens SS. On the psychophysical law. Psychological Review. 1957;**64**(3): 153-181. DOI: 10.1037/h0046162

[28] Michell J. Stevens's theory of scales of measurement and its place in modern psychology. Australian Journal of Psychology. 2002;**54**(2):99-104

[29] Vinberg ĖB. A course in Algebra. American Mathematical Society. 2003; **56**:511

[30] Bond TG, Christine MF. Applying the Rasch Model: Fundamental Measurement in the Human Sciences. 2nd ed. New York: Routledge; 2020. p. 348

[31] Tennant A, McKenna SP, Hagell P. Application of Rasch analysis in the development and application of quality of life instruments. Value in Health. 2004;**7**:S22-S26

Chapter 6

Semiotics and Decision Making Using AHP in Medicine

Gheorghe Jurj

Abstract

This chapter discusses the intricate relation of medical semiotics and medical decision making. From the very beginning to arrive in a medical context, some signs have to be perceived by the person him/herself, by others, or by using some instruments. If these show a deviation from the corresponding normative sign, the process of collecting signs and necessity for a decision making is proximate. We discuss some types of medical signs directly related with the decision-making process, for example, normative signs, intra- and inter-subjective signs, subjective and objective or instrumental signs, scalar signs, etc. These signs create configurations of signs, and configurations of signs are the fundament of medical and decision rationing. The model of Analytic Hierarchy Process (AHP) uses criteria that are scaled for decisions among different options, but in medicine to set criteria and scale values is a semiotic process because it refers signs. An example of AHP decision making was given for a domestic medical decision about addressability, that is, going to the medical institutions or not, but AHP can be used, and already was, in most of medical decision realm, from addressability to diagnosis, treatment, follow-up, public health policies, or strategies.

Keywords: medical signs, normative signs, medical semiotics, medical decision making, analytic hierarchy process

1. Introduction

One of, if not the most important problem in medicine, is finding grounds for therapeutic action. Any physician-patient relationship, as well as the whole field of practical or community medicine, aims for action, and action in medicine follows *decision making* (DM). For its part, DM must be based on a set of data that are entirely of the nature of signs, whether they are subjective (what a person feels) or objective (what can be observed from the outside), or instrumental, through paraclinical means. Before introducing these signs into an evaluation system leading first to a diagnosis and then to therapeutics, they must be defined. The discipline that deals with the definition and discussion of signs is semiotics. Semiotic accuracy is the first step towards a correct medical decision. The notion of "health", together with the other associated concepts (disease, illness, well-being, quality of life, etc.), defines a whole *semiotic realm*, which was and is conditioned both historically and culturally and without which medicine cannot work. Thus, the two fields, semiotics and decision making are *sine qua non* conditions within any medical practice, permanently involved at every level of evaluation and action, often intricate and even difficult to distinguish at first glance (the diagnosis itself is a decision based on signs to which a certain value is assigned).

In this chapter, I first describe diverse types of signs used in medical context (explaining the difference with pathological signs, related to diseases, subchapters 1 and 5) and their significance for the DM process in a point where Semiotics and Analytic Hierarchy Process (AHP) converge: the scales for degree descriptions of signs (subchapter 2). I explained in a simple example how an AHP can model the first issue of medicine, that is, addressability (analysis made in subchapter 3), and then how qualifying the signs with different determinations can increase their values in formatting reliable criteria for AHP (**Figures 1–4**). In the 6th subchapter, we discuss a model about how successive configurations of signs lead to successive layers of decisions related to all stages of medical reasoning and acting (see **Figures 5** and **6**), and in the last one, a brief description of domains where AHP (a general model described by **Figures 3** and 7) was already used in medicine and most probably will be used extensively in the future, beginning with patients' decisions to public health policies. A general flowchart of this chapter methodology is illustrated by **Figure 8**.

1.1 Signs in a medical context and pathological signs

To arrive at a diagnosis, the medical experts (physicians) must specify a number of signs that they learned in medical school, the so-called/disease signs/ or / pathological signs/. Pathological signs are those signs which the patient presents that represent deviations from/the normal/ and through which the physician can potentially define a possible disease. The discipline that studies such signs related to possible diseases is academically called *medical semiology*. In what follows, we will not discuss this discipline as defined in medical textbooks. We will rather focus on its theoretical foundations from the broader perspective of medical semiotics [1, 2], and semiotics as C. S. Pierce [3], C. W. Morris [4], U. Eco [5], and others [6] defined it.

From the very beginning, it must be stated that not all the signs that a physician takes into account are necessarily pathological signs. Certain signs that may not be directly correlated with a disease (habits, hygiene, mentalities, work stress, and even the person's temper or behavior), but which signal a deviation from a state considered by the patient or the physician as /normal/, can still become *signs in a medical context*.

The *medical context* is that context in which details of the private or the public life of a person or a community are connected to a possible impairment of health or well-being. A conjunctural sadness may not necessarily be a feeling connoted as having pathological significance, but when the sadness is prolonged or when it becomes a possible cause of a psychological (depression, burn out syndrome) or immunological disorder, it becomes significant in a medical context. That is why signs in a medical context could be defined as any signs that can lead to or show a disturbance in the person's state of normality, regardless of their nature. Sometimes, these signs are not immediately connoted as pathological signs, rather as the "way of being" of the person, but even this way of being can be the source of possibly pathological deviations. For example, a person defines themselves as follows: "I get angry very easily, sometimes I become violent verbally or physically ... " that's how I am, there's nothing I can do about it"; it is a state of irritability, nervousness that they consider normal for themselves, but when it is at the origin of behavior disorders, we already have a deviation from normality, which often requires medical intervention (psychotherapy, behavioral therapy, medication) following a DM within the medical spectrum. A certain gene that provides information for making a protein on the surface of the white blood cells, called the HLA B27 complex, can be present normally in some people, but in the context in which they present joint or spine pain, or digestive disorders, this presence becomes indicative of a group of diseases called Spondyloarthropathies or the Inflammatory Bowel Disease (IBD), although it is part of the person's genetically determined "way of being." The same is the case with all genetic disorders or genetic predispositions: in these cases, we have a deviation from a statistical /normality/, compared to other people, but the person in question does not have a deviation from themselves.

In this instance, I am referring to the medical context as a part of the medical communication structure, which aims to convey medical signs—from transmitters to receivers, through transmission channels, according to R. Jakobson theory of communication [7]—and possibly help to make a decision. It is defined every time there is communication in which a potential medical decision could be made, hence its pragmatic dimension.

2. Normality as a /normative sign/ with reference value

The most typical example of a medical context is that of the physician-patient relationship, but before reaching this phase, there are some previous stages. In the simplest terms, it all starts when a person "doesn't feel good anymore," "doesn't feel able to do something," or simply "feels that something is wrong." Other times, the person does not feel anything, but the laboratory analyses are deviated from /normal/, or those around notice a change in their state, behavior, abilities, and make the assumption that something has changed, and this something is of a medical nature. In other words, the beginning of the medical context is a deviation from a state implicitly or explicitly assumed to be that of /normality/. Normality can be defined in many ways, the simplest and the vaguest definition is that of World Health Organization (WHO): "*health is not merely the absence of disease or infirmity but a state of complete physical, mental, and social well-being*" [8]. What is of interest here is not only its denotative value as a generic signs/health/, but the deviation from it felt subjectively (displeasure, pain, helplessness, etc.), objectively apparent (changes in appearance, functionality) or revealed only by paraclinical instruments (laboratory, imagistic). As G Canguilhem [9] and M. Foucault [10] stated the concept of health or disease is determined not so much by science as by the idea of cultural determined normativity, "*but normality has no consensual definition in medical literature*" [11–13].

Every time a medical context appears, the signs presented, either by themselves or by others (family or physician), are compared explicitly or implicitly with what is considered to be /normal/. But normality itself is constituted by signs, which are mental or social representations of normality. These signs that explicitly or implicitly define normality and that are compared to the patient's signs each time, one by one, are *normative signs* [14]. In the terms of C. S. Pierce's semiotics, their *referent or object* is that which is considered normal, *their representation* is the way they express themselves (e.g., the color of the face or the acceptable range of an analysis), and their *interpretant* is either a person or a system in which they function as values of reference [15]. Without normative signs, medicine would not be possible because, at each stage of the evaluation in a medical context (whether it is about addressability, diagnosis, or treatment), what is ab-normal or dys-functional is defined in relation to them. Punctually, each normative sign becomes the *center of a semiotic field*, or the reference value in which the actual signs discovered in real patients fit [16].

2.1 Normative signs are qualified

If we take an implicit normative sign, for example, /normal hair/, around it we can describe variations in which qualifications are implied. /Normal pubic hair/ is normative to the adult age, but if it appears as /precocious pubic hair/ we have a variation from the norm and the possibility of some hormonal disturbances such as too early puberty: here the qualification of the normative sign is age. If an adult male presents /loss of hair/, this situation can be sometimes considered /normal/, if it is an inherited hormonal family feature: this is a gender qualification of /normal hair/. But if this condition is disturbing for the person, they can request treatment or even implantation, and this is the reverse process, whereby entering a medical context (address a physician for implantation) transforms a trivial personal feature into a sign that demands choice in the DM process. If the /falling of hair/ is a deviation from /normal hair/, a physician has to qualify the new sign and, according to certain determinants, decide if it is pathological or not, or if it is the object of a medical DM. For example, this sign can be qualified by (a) *Age:* it is closer to normal in older age; (b) *Physiological situation*, for example, after pregnancy and long breastfeeding; (c) *Medical condition,* if it appeared after chemotherapy; in this case even if the / falling of hair/ is non-normal, it is not considered pathological in itself, but rather an / iatrogenic sign/ induced by therapy, and as such it is not directly connected with a specific DM; (d) *Type*, if it appears in small spots, the so-called *alopecia areata* (if it is total, *alopecia totalis*), and it implies not just the scalp hair but also axillar, pubic hair, and eyebrows.

What we generally call laboratory findings, all refer to certain normative values, and can be directly related with a DM. Their general model is a certain interval where the measured parameters are being considered normative, so the normative sign **/S/N** ranges from values **X** to values **Y**. What is lower than that value is considered hypofunction or **/S/**– values; what is above are hyperfunctions or **/S/+** values; the values increase numerically, according to the arrow (**Figure 1**).

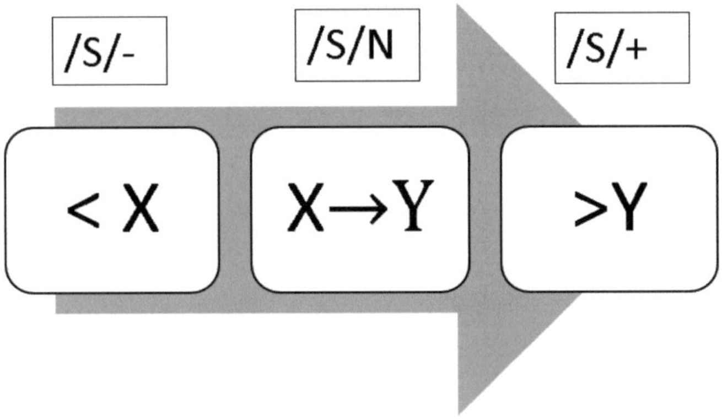

Figure 1.
Range of values as normative sign against which non-normal variations are established.

Many parameters are described by just one interval as **/S/N** = **X** → **Y**, so, apparently, an interpretation of this type of signs is rather simple. However, for certain substances, the normative sign /S/N is qualified by *physiological determinants* (e.g., the blood sugar interval 90–110 mg/Dl is normative just in the morning, before eating). Therefore, instead of having just one normative interval we will have more, qualified by determinants. For example, when taking into consideration the so-called ~hormonal constellation~ in women, there is a set of four hormones: Estrogens, Progesterone, FSH (Folliculin Stimulating Hormone), and LH (Luteinising Hormone). The normative values for these are dependent both on age and the period of menstrual cycle. There are five types of normativity: (a) Before first menstruation; (b) In the folliculinic phase of the cycle—therefore making the sample is recommended in day 3–4; (c) During ovulation; (d) In the progesterone phase of the cycle—recommended in day 22–23; e) After menopause.

Now, the normative sign becomes */normative sign in certain conditions/* and these conditions are classified according to certain criteria. If the normative sign is properly classified, it will function as a reference for the comparison of the actual sign of the person (patient) and it will become a trustful factor for a medical DM.

2.2 Types of scales for normative signs

Scaling is an important process in Analytic Hierarchy Process (AHP) for DM as stated by Thomas L. Saaty [17]. To decide whether a sign is pathological or not in a medical context, and hence its use for a AHP in DM, several types of scales are being used, in which the normative sign can be located:

a. *at the beginning* of the scale ("my head doesn't hurt" as beginning for pain scale) and the pain scale goes up;

b. *in the middle* of the scale (e.g., laboratory values). From there, the scale goes up and down: the values are very low—lower—*/normal/*—higher—very high;

c. *at the end* of the scale ("I feel tired"), where the scale starts from the normative sign "normal energy" and goes downwards.

The scaling expression is essential for each of the DM moments, because it can lead to one decision or another depending on the relative importance given to that sign; in other words, *the gradation of the signs is the expression of their assigned value*. There are several grading possibilities, which are used daily in any medical context, both by people with different conditions and by physicians, as shown by E. Cirino [18]:

1. The type of scaling through easily definable categories, Categorical Scale (CS), in degrees: not at all—a little—medium—a lot—very much—worst possible.

   Categorical scales (CS):

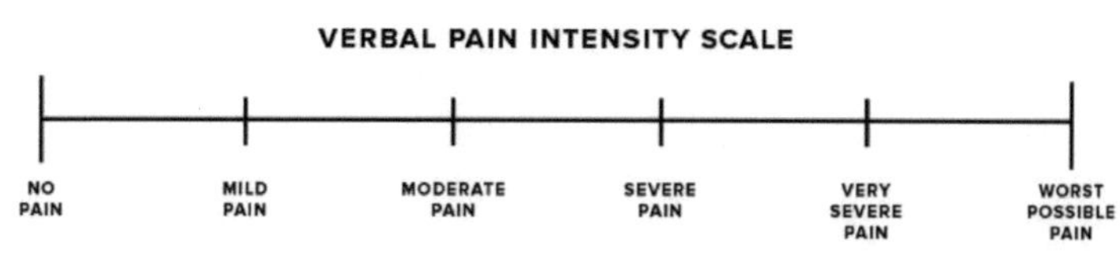

For children and for some patients, the representation may be *iconic*.

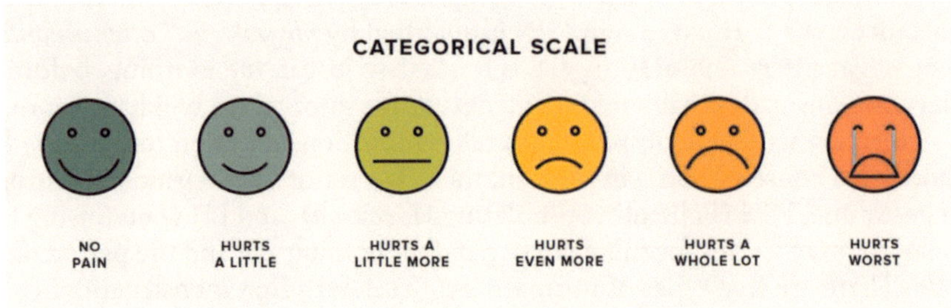

2. In the form of numerical values, Numeric Rating Scales (NRS), for example, define pain (or worsening, improvement after treatment) from 0 to 10.

   Numerical rating scales (NRS):

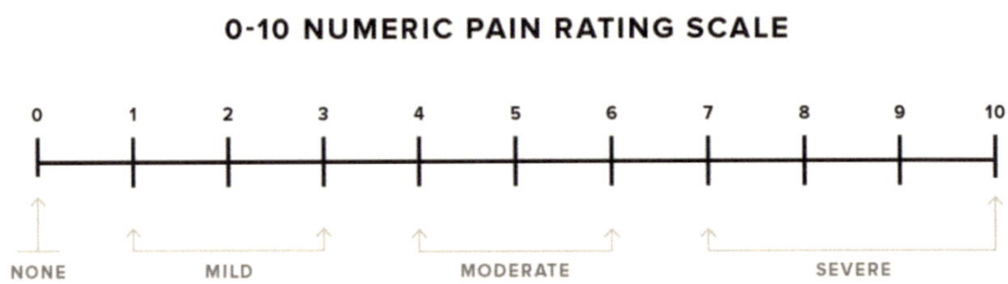

3. Of the type of percentages: "what percentage is better/worse", widely used for the subjective evaluation of improvements or worsening after a treatment ("I feel ~50% better"), but also for pain as in the Visual Analog type Scales (VAS)

   Visual analog scale (VAS):

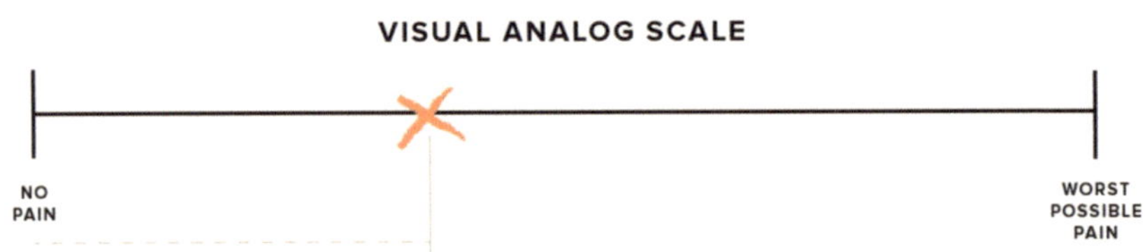

4. The qualitative type, for example, pale-yellow-normal-congestive-cyanotic. In the case of qualitative signs, for example, the color of the face, the scaling is more difficult. In this case, the normative sign is in the center of the semiotic field which it generates. When the physician sees the color of a patient's face as a sign and interprets it as /pale/ or /congestive/, the normative sign /normal color/ is implicit. Around this sign, a whole *contextual semiotic field* unfolds, wherein variations of nuance may represent medical signs [16].

   For the formalization in AHP, a scaling of the criteria represented by these signs is needed. The first three types of grading are easily translated into a system from 1 to 9, with odd "strong" values.

   - "not at all," 1.0% = AHP scaling 1.

- "a little," from 1 to 3, up to 30% = AHP scaling 3,
- "medium," from 3 to 5, up to 50% = AHP scaling 5.
- "a lot," up to 7, up to 75% = AHP scaling 7.
- "very," up to 10, or 100% = AHP scaling 9.

2.3 The physician-patient relationship as a semiotic process in the DM perspective

When presenting to the physician, the person is subjected to an interrogation, clinical examination, laboratory tests or imaging, so that the physician can make a decision. The decision involves three aspects: (1) what is it based on? (2) what diagnosis does the patient have? (3) what treatment is to be prescribed (prescription of drugs, recommendation of lifestyle modification, prohibitions, or even surgery, etc.). Throughout the whole process, from the feeling that "something is wrong" to a medical action, everything involves signs. The patient perceives their condition as a sign of something, they communicate the condition to those around them or to the physician, the physician adopts diagnoses or treatments, and the final evaluation, if the medical decision was correct, is also made through signs.

At every moment of this continuous semiosis, every semiotic level is accompanied by a decision. From the first moment of the patient's presentation to the physician, an interpersonal relationship is established between the two that is neither neutral nor generic, but has a cultural predetermination that involves all kinds of expectations when entering into such a relationship. The physician-patient relationship, right from its inception, has a certain meaning for both. In semiotic terms, any patient in front of a physician is a source of signs embedded in the generic sign of /patient/. Conversely, the patient perceives the person in front of them under the sign /physician/, who in this context is not just a human being in general, but is coded as /physician/ (given the place, gown, stethoscope, hospital). Each of the two parties is involved in a communication relationship as senders and receivers of messages in the form of signs. According to Roman Jakobson's model, any communication relationship involves a sender, a receiver, a message, code transmission channels, and a context (**Figure 2**).

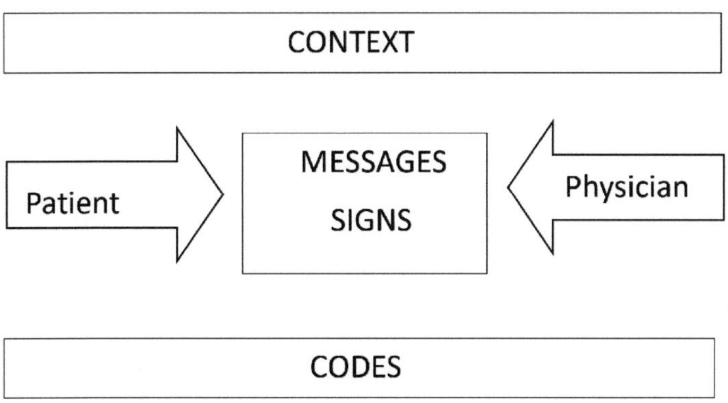

Figure 2.
Patient-physician communication adapted from R. Jakobson model.

Between the patient and the physician, what is exchanged are messages: not unidirectionally, rather in both directions (e.g., the patient says what hurts, the physician explains what it could be). These messages are of the nature of the signs themselves, the interview revealing a twofold possible centering that should be integrated [19].

In this communicative ensemble, the context is particular, defined on the coordinates of the so-called "medical consultation," whose purpose is to issue a decision regarding the patient's state of health. This context is from the very beginning oriented by specific concepts that permeate the consultation framework, such as for example that of "diagnosis" and "what needs to be done." In other words, the *meaning of the signs is pragmatically conditioned by their use* and the perspective of a DM. On the other hand, this communication has specific codes, from the coding that appears through space, the relative position of the two, the dress code of the physician, to verbal codes (the specific language of both, which are controlled by the code) or psychological (the patient knows they have to answer questions because they encoded the assumption that the physician will help them). We identified several types of codes in the medical context: descriptive, associative, prescriptive, and permissive codes [20].

3. Domestic medical DM, the issue of addressability

Before a person goes to the physician when certain signs suggest that a possible medical problem has arisen, there is always the question of a decision making: "what do we do, stay home or go to the physician?" It is the issue of addressability: deciding between domestic medical action or addressing a physician or a medical institution.

3.1 "The child has fever" and the AHP modeling in DM

When the notification of non-normality is made by others, for example a mother who notices that the child has a fever (or a social group that notices a violent behavior), the supposedly non-normal person is the object of perception from the outside, and the judgment of their condition is a reporting to the idea of normativity. It is a process of comparison between two categories of signs: (a) signs by which the state of that disturbed person is perceived (for example, the red face of the child and the high temperature of the forehead felt on the hand by the mother) and (b) signs that represent normality, normative signs (the mother tells herself "before, the child was normal"). From the comparison between the two strings of signs, the non-ordinary, non-normal state is inferred, and from here the path to the decision-making process is opened.

Any referral to a physician requires that the potential patient or people around them make the decision. Although the process seems simple and is usually undertaken without full awareness of the decision moments, it could be modeled because it essentially includes all the stages that can give rise to a Hierarchical Analysis Process (AHP). Generally speaking, in any AHP there is a specific structure, where T = *Task*; O = *Options*, O1, O2, O3, O4 and C = *Criteria* C1, C2, C3, C4.

Thus, for example, in a context where the child has a fever, the mother is faced with the following options (O): (O1) *to do nothing* and wait for it to pass by itself; (O2) to start *self-medication* (e.g., anti-thermic); (O3) to contact the *family physician*; (O4) to go to the Em*ergency Room*.

What would be the criteria (C) that the mother implicitly or explicitly takes into account to make this decision? (1) C1 = *cost* in time, energy of mother or money; (2) C2 = *risks*, the uncertainty of the diagnosis (is it a simple virus, or it hides a complicated diagnosis); (3) C3 = *aggravation* in time or amelioration; (4) C4 = *safety*, accessibility of the physician or the feeling of being safe of the mother towards her decision about child. If we take into account the four options for action and the four criteria, an initial scheme of an AHP can be made. The APH structure of this DM problem ("the child has a fever - what do I do?") is as follows, yet the number of criteria or options can be variable from case to case and from problem to problem (**Figure 3**).

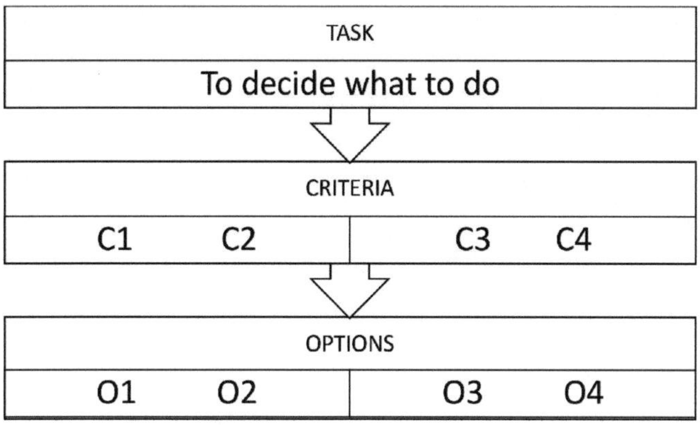

Figure 3.
Construction of a decisional hierarchy with 4 criteria and 4 options.

According to AHP, the criteria will each be scaled from 1 to 9, where 1, 3, 5, 7, 9 represent degrees of importance assigned by the decision maker: 1 = unimportant, 3 = a little important, 5 = medium importance, 7 = high importance, 9 = very high importance. The even numbers 2, 4, 6, 8 = represent transition values between the odd degrees. Here beneath an example how this problem can be formulated and solved in terms of AHP, using the model for medicine conceived by Dolan et all [21, 22]:

- Step 1. Setup the hierarchy

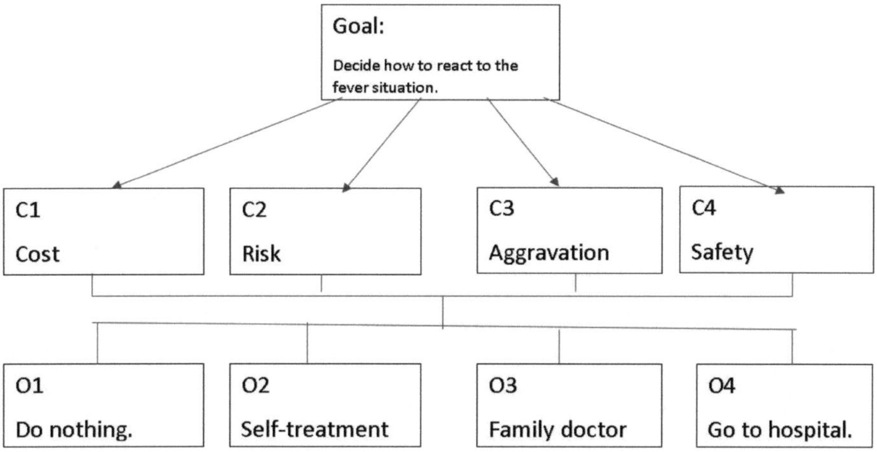

- Step 2. Pairwise comparisons.

A. Comparison matrix of the 4 criteria vs. goal.

Criteria vs. goal	MC	MR	MA	MS	Weights
Minimize cost (MC)	1	1/3	1/4	1/2	0.15
Minimize risk (MR)	3	1	1/2	1	0.39
Minimize aggravation (MA)	4	2	1	2	0.26
Maximize safety (MS)	½	1	1/2	1	0.18

B. Comparison matrix of the 4 alternatives vs. MC.

Minimize cost	DN	ST	FD	H	Weights
Do nothing (DN)	1	3/2	2	5/2	0.37
Self-treatment (ST)	2/3	1	5/3	2	0.28
Family Doctor (FD)	½	3/5	1	3/2	0.19
Hospital (H)	2/5	½	2/3	1	0.13

C. Comparison matrix of the 4 alternatives vs. MR.

Minimize risk	DN	ST	FD	H	Weights
DN	1	2/3	1/3	1/3	0.13
SD	3/2	1	1/2	2/3	0.19
FD	3	2	1	3/2	0.39
H	3	3/2	2/3	1	0.29

D. Comparison matrix of the 4 alternatives vs. MA.

Minimize aggravation	DN	SD	FD	H	Weights
DN	1	2/3	1/2	1/2	0.13
SD	3/2	1	1/2	1/3	0.17
FD	2	2	1	2/3	0.3
H	2	3	/2	1	0.39

E. Comparison matrix of the 4 alternatives vs. MS.

Maximize safety	DN	SD	FD	H	Weights
DN	1	3/4	1/3	1/2	0.13
SD	4/3	1	1/2	2/3	0.17

Maximize safety	DN	SD	FD	H	Weights
FD	3	2	1	2	0.44
H	2	3/2	1/2	1	0.25

- Step 3. Weighing the decision elements.

The weights were determined using the method of normalized row sums. That is, each weight was computed by summing the values of the corresponding row and then divided by the sum of the elements in the matrix. This gives an acceptable approximation of the normalized eigenvector of the comparison matrix.

- Step 4. Synthesis.

The following table captures the information distilled thus far:

	Criterion			
	Minimize Cost	Minimize Risk	Minimize Aggravation	Maximize Safety
Alternative	0.15	0.39	0.26	0.18
DN	0.37	0.13	0.13	0.13
ST	0.29	0.19	0.17	0.17
FD	0.2	0.39	0.3	0.44
H	0.13	0.29	0.39	0.25

From this table we read:

$$DN = 0.37 * 0.15 + 0.13 * 0.39 + 0.13 * 0.26 + 0.13 * 0.18 = 0.16$$
$$ST = 0.28 * 0.15 + 0.19 * 0.39 + 0.17 * 0.26 + 0.17 * 0.18 = 0.19$$
$$FD = 0.19 * 0.15 + 0.19 * 0.39 + 0.3 * 0.26 + 0.44 * 0.18 = 0.33$$
$$H = 0.13 * 0.15 + 0.29 * 0.39 + 0.39 * 0.26 + 0.25 * 0.18 = 0.27$$

In conclusion, the final ordering of our options is:

$$\textit{Family doctor} > \text{Hospital} > \text{Self} - \text{treatment} > \text{Do nothing}.$$

Of course, the assignment of priority numbers and the whole AHP mathematical process will not be done by a mother with a feverish child (who may not know matrix math and never heard of eigenvector), but essentially, she makes the decision unconsciously, while prioritizing in fully unexpressed and unformalized scales, according to her judgments.

3.2 Addressability as a decision-making problem

Decision models like the one discussed above can fit into the broader category of DM that has as its object the *addressability* to a physician or a medical institution, and represents in itself a medical, but domestic decision, in which one chooses between several options. This is one of the serious problems in medicine because, on the one hand, an excess of addressability unnecessarily burdens the medical system with things that could be solved domestically, but, on the other hand, the lack or delay of addressability can lead to the serious evolution of some diseases with complications. The two extreme situations were both present during the COVID-19 pandemic and both proved their disadvantages and consequences [23]. Many other examples can be given:

- The lack or scarcity of addressability in particular communities: remote, or poor, or lacking medical institutions. In these cases, although there might be an addressability decision, achieving it is difficult and may involve insurmountable logistical problems.

- Lack of addressability for certain social categories, for example, in less educated, poor or middle-class environments under the influence of ideological or religious factors.

- The lack of addressability that appears through previous experiences interpreted as unpleasant: long waiting time for consultation with the GP or specialist, the experience of unsatisfactory or superficial treatment felt by the patient in some situations in the past (sometimes precisely through the overloading of the GP or the emergency system).

- Lack of addressability for personal reasons, of a psychological nature (e.g., "fear of the physician," or mistrust of the physician/medicine).

- or the nature of some personal beliefs (e.g., neglecting certain symptoms "for which I don't think I need to go to the physician").

4. From sensation to the medical sign—intrasubjective and intersubjective semiotic processes

The moment of emergence of a medical context is that of a deviation from what the patient or those in their immediate social environment consider to be normality. In the patient's language, they can be expressed verbally ("I don't feel well," "it hurts," "I'm sick," "I cannot," etc.) or non-verbally (through changes in color, dynamics, behavior).

A first level of meaning, *the intra-subjective* one, is that in which the person realizes that, in one way or another, they have moved away from their previous state. This realization occurs as a result of an immediate fact, which can be a sensation or a certain state of fact of the nature of Firstness, in Pierce's terminology [24, 25]. This fact is imposed on the patient as something more or less defined, which they enter a relationship with (of the nature of Secondness), while entering a relationship with themselves comes to be interpreted also as a disturbed state, or as pain, following an act of mental representation that functions as an interpretant of the nature of Thirdness. Expressing it is the next step, through which this process becomes presentable to others (words) or to themselves (groans).

Thus, we have signs that follow each other and that generate other signs at each step: (a) a first chain is *intra-subjective*, from sensation, to perception, to mental representation, as "not good," "discomfort," "illness"; (b) from this mental representation sign to its expression in other communication systems, the transition from the intrasubjective level to the *inter-subjective* level of signification, in which all possible categories of signs (indexes, icons, or symbols) can be used.

For example, when a child puts his hand on his abdomen and squirms because of a tummy ache, the sign is of the nature of the *index*. When a throbbing pain is signified by a gesture such as clenching and unclenching the fist, the representation is *iconic*, an image of the type of pain which is felt. All verbal expressions are *symbolic* representations and appeal to a conventional language to translate a particular sensation [26]. It is worth noting that when dealing with *verbal signs* there is most often an *intentionality* of the sufferer. Sometimes, however, *para-verbal* emissions (e.g., moans of pain or interjections) are merely indexes and not symbols and indicate the referent directly, from the physiological fact to its *non-intentional* expression.

Non-verbal expression can lead to signs that are transmitted through any channel: visual (gestures indicating pain or sensation), auditory (sighing, moaning, sound of voice), olfactory (smell of sweat or breath). Non-verbal signs can in their turn be intentional, for example, they indicate the location of a pain, or non-intentional, for example a hunched position during colics. These unintentional signs express disturbances at the organic level, for example, an intense spasm that compels yawning. Intentional and non-intentional signs, as well as verbal and non-verbal signs, can be associated, so that repetitions of messages with the same referent appear, which increase their semiotic value for the decision-making act. If the child moans, always puts his hand to his ear and says "it hurts", this *complex of different types of signs transmitted through different channels but having the same referent* makes the mother or the physician take the child's condition seriously, to think of otitis and be on the verge of making a decision.

We can summarize the appearance of these signs through a sequence of stages from sensation to expression, as in the **Figure 4** (transformed from [24]):

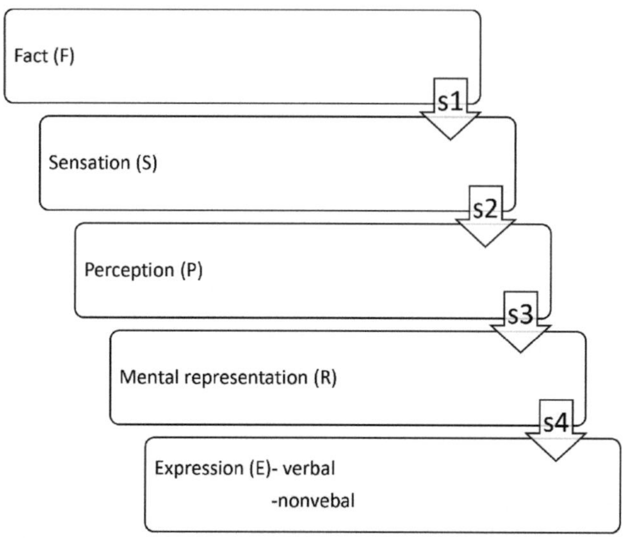

Figure 4.
Succession of signs from facts to expression.

In the conception of C.S. Pierce, any sign has a triadic structure, in which *the Referent* (or the object of the sign) is what is referred to, *the Representamen* (or the sign itself) is what the referent represents, and *the Interpretant* is the one through which the representation is correlated as being of the object [3, 27]. The primary intrasubjective referent is an undefined sensation caused by a fact or by physiological change and the first sign is established when this sensation becomes a perception that represents the sensation (something is perceived as /painful/). The second semiotic level is that of the perception that has become a referent which causes a mental representation to be defined (something is perceived as/burning pain/). The third semiotic level is the one in which this representation is expressed verbally ("my head hurts") or non-verbally (I point my finger to my temple, meaning that it hurts there). An expression of this course is:

$$F \rightarrow /S/(f) \rightarrow /P/(s) \rightarrow /R/(p) \rightarrow /E/(r), \tag{1}$$

in which each semiotic level is an interpretant for the previous one and becomes an object for the successor. In this way, there is, on the one hand, *the preservation of the original referent* along the entire chain, and on the other hand, the need for an *interpretation*, sometimes possibly even distortion, of the message at each transition from one level to another.

This chain $F \rightarrow S \rightarrow P \rightarrow R \rightarrow E$ is conditioned at each stage:

- $F \rightarrow S$, like $S \rightarrow P$, are *indexical* or direct signs, but the semiotic process can be interrupted. For example, if the nervous system is embodied by a peripheral paralysis, or the patient is unconscious, S no longer passes into P, although the physiological fact F (burning or nerve irritation) exists.

- The mental representations of some $P \rightarrow R$ perceptions are *iconic* signs, already conditioned not only by what is felt as a perception, but also by the mental interpretation by which that sensation is represented (pain is represented "as if ... it is burning," "as if ... it is a stich," "as pressure").

- The expressions from $R \rightarrow E$ are *symbolic* signs, which already need certain expression codes. These can be clearly established by the convention of the spoken language in a certain language ("it hurts," "mă doare," "ça fait mal") or, as it happens in the case of certain expressions of pain established by unconscious cultural conventions ("oiii," "auuuu," "auhhh").

- Some non-verbal expressions such as the posture during pain (e.g., /bend double/), the type of walking (/stumble walking/), the color of the face or palms (/pale face/, /yellow palms/) can be direct expressions of certain physiopathological facts (pain, paralysis, jaundice) and are iconic signs; the referent is that fact and the expression is no longer subjectively mediated by S, P, R, but a direct relationship $F \rightarrow E$ is established.

- In certain psychosomatic conditions, we can focus on the mental representation (R) that is expressed (E) toward oneself or others even in the absence of a determining F physiological or pathological fact. This retrograde path, from the expression to the representation of sensations or perceptions, depends on one's own production of mental signs that build their own representation, sometimes up to the reverse reconstruction and the appearance of sensations or perceptions (a simple example is a fever with muscle pain that appears after anger or sadness).

4.1 "A burn," signs as criteria directing to DM

The question that arises is whether this semiotic chain has any importance for clinical medicine and the process leading to DM? Is the process of signifying simple facts of this kind somehow related to the decision process, or in other words, how are *the two processes interconnected* in a medical context? Let us take a simple example, a burn at home, and examine several situations as possible options for action, decisions, and addressability in simple "if-then" decision terms, where O are options:

1. A 1st degree burn with local redness leads to options O1 = do nothing, or O2 = apply ice.

2. A 2nd degree burn over a larger area leads to O2 or further to apply an O3 burn spray.

3. A 3rd degree burn (bullae), on some surface, leads to O2, O3, and possibly O4 = an analgesic drug and to O5 = local dressing.

4. A 3rd degree burn with large extension can determine either O1 → O5 or directly O6 = presenting to the emergency room.

5. A 3rd degree burn with a large area can determine O7, calling the ambulance, where a specialized treatment will begin.

For all these options, the decision mechanism is one: Fact → Decision → Option, wherein the criteria (C) are not very elaborate, but are based on objective (what is seen) and subjective (what is felt) signs that are relatively easy to understand: C1 = degree of burn, C2 = its surface, C3 = pain, C4 = anxiety about pain or complications. The scaling of the criteria is in large steps, of the type "a little- normal -a lot" for the degree, extent, or intensity of the pain. This way, the decision variants are according to the objective or subjective scaling of the signs and we can design a simple map of AHP.

It is worth noting that in the semiotic chain from F to E, the intensity of each of the component elements can be scaled either spontaneously (as it happens in a burn) or according to easily decipherable scales (degree of burn, surface), so that each sign is associated a degree, which will eventually become a criterion and a decision factor. For example, the pain can be relatively easily interpreted from 0 to 9 or in simple forms such as weak, medium, intense, excruciating. In this interpretation, any of the elements of the semiotic chain can be decisive for making a decision.

5. Types of medical signs

In medical DM processes, we always operate with signs of different types and origins. They can be synthesized, from a semiotic point of view, in several categories [28].

5.1 Subjective signs, objective signs, instrumental signs

A first and classic classification of medical signs was made according to their field of provenance or source. *Subjective signs*, commonly called *symptoms*, represent a subjective experience that cannot be identified by anyone else, its referent is about how the patient feels, reacts, or behaves in certain conditions, and the means to obtain them is mostly by their being spoken. *Objective signs* (sometimes called simply signs, in medical language) are observable phenomena that can be identified by another person, be it non-medical (e.g., family) or a medical person. Their domain is very vast, and they can be: (a) *static*, resulting from the appearance of some parts of the body (color, swelling, dryness, cracks, symmetry, etc., of eyes, skin, tongue, limbs), to the specific appearance of some injuries (rash, eruptions, ulcers), or (b) *dynamic*: walking, breathing movements, movements of parts, speech as phonation, voice sound, etc.

The *instrumental signs* are those for which an external source, device, or instrument is used. Among the first signs of this kind were those that amplified or clarified the ordinary senses with simple instruments: stethoscope, otoscope, ophthalmoscope, dermatoscope, microscope. In parallel with the scientific and the technological development, three other large categories of instrumental signs appeared: (a) *laboratory signs*, which are obtained through specific analyses of body fluids (blood, urine, cerebrospinal fluid, pleural exudate, etc.) or other products (stool, stones, sputum, etc.); (b) *imagistic signs*, in which iconic images of parts of the body are formed by means of instruments based on physical phenomena. The first type was the Roentgen machine, but now this field has become extremely broad (Magnetic Nuclear Resonance, ultrasound, Computer Tomography, etc.); and (c) *graphic signs*, wherein the instruments translate a bodily function into a graph, with symbolic signs (Electrocardiography, Tympanography, Electromyography, etc.).

5.2 Simple signs—qualified signs

Simple signs are those that generally signal a fact from any of the above categories. These signs often have a simple index value, e.g., subjective /pain in the wrist/, or objective/swelling of the wrist joint/, or instrumental /temperature of 39 degrees Celsius/. For the patient, they can be at the origin of a DM related to *domestic treatment* (H) or *addressability* (A). It should be noted that instrumental signs (laboratory, graphic, or imagistic) are actually never simple signs, because they require an interpretation of the entire generated semiotic field. For example, the image of a formation will be qualified by size, location, relationships, intensity of the image (e.g., on ultrasound transparent for a cyst, or hyperechogenic for a tumor).

In making a diagnostic decision, these simple signs are of little semiotic, and implicitly decisional value. Therefore, through medical questions and investigations related to professional skills, the physician will try to find out as many details as possible about these signs; in other words, they will try to add them qualities, *to qualify* them [29]. Thus /wrist pain/ must be qualified:

- *In context* (after what it appeared? After a stress, cold, other illness, or other treatment?);
- *In time* (when is it? Day-night, morning-evening, summer-winter?);
- *In quality* (how is it? sharp, weak, burning, stitching, excruciating?);

- *In localization* (where is it? Just a fist-to-both, does it radiate to the fingers or elbow?);
- *In modalities* (how does it get worse or better? Hot – cold, movement – rest?);
- *In association* (with what is it associated? With restlessness, or fever, or vomiting?).

This process is extensively use in homeopathy, where the qualifications may lead to a DM for homeopathic remedy as treatment [16].

The qualifications by objective signs are being added (the hand is very swollen, hot, he cannot move it). And after the DM to do investigations, laboratory signs (high Blood Sedimentation Rate, high level of uric acid, high level of fibrinogen) or imaging (signs of inflammation of the metacarpal joints) may appear as necessary for a diagnosis DM. Thus, qualified and determined signs on all directions of medical investigation, when put together, better define the diagnostic criteria and implicitly a diagnostic DM.

5.3 Present signs and historical signs

Historical signs are those reported by the patient about personal or family illnesses or result from medical records of the patient. Although they are apparently not related to the actual main complaint, they can qualify actual signs. For example, the patient remembers that they had episodes of wrist pain several times, in the cold or during the holidays (when they ate more pork, sausages, smoked meat), but they ignored them, as well as the fact that the patient's father and family had gout.

Historical signs can be *personal* (e.g., a patient had recurrent purulent tonsillitis as a child and now he complains of strong joints pain) or they may run in the *family* (e.g., a patient with a mammary lump has a mother operated for breast cancer). If the actual signs define a *synchronic consideration* of the current signs, the historical signs define a *diachronic perspective*, wherein the past and the present signs are connected. This connection can be decisive for scaling certain criteria for the diagnostic DM.

5.4 Scalable signs and iconic signs

As we have seen, most of the signs that are considered in a DM can be scaled. The subjective signs are being scaled by intensity scales, the laboratory signs by the normality range against which everything plus or minus can be scaled (e.g., hemoglobin - /normal / or slightly low, very low, increased, very increased). However, some pictorial or graphic signs are difficult to scale. (For example, the presence or absence of a kidney stone on urography or the presence of fibrillation on EKG). These are iconic signs directly related to a certain pathology, and although they reach the physician when mediated by the appropriate instrument, they have an "either-or" meaning. In other words, they are *eliminative signs*, high-graded criteria that can exceed all the others.

Other signs are characteristic of only one disease; they are also called *pathognomonic signs*, and they have the highest specificity and high positive predictive value (e.g., Koplik's spots in measles). Most often they are found by the physician simply because they look for them in a certain configuration of other signs.

6. Semiotic and decision chains in medicine

The stages of the entanglement of semiotics with those of the medical DM can be formalized. If we make the following observations:

/S(s)/—subjective signs, intra- and intersubjective (from 1 to n—as many as collected).
/S(o)/—objective signs (from 1 to n).
/S(i)/—instrumental signs (from 1 to n).
/S/ means that it is about signs.
Cf /S1/—Configuration of signs.
O = possible options or alternatives.
DM 1 = domestic care or addressability.
DM2 = diagnosis, DM3 = treatment etc.

We can formalize the relationship between them, so that we consider the first configuration of signs Cf/S1) as the qualified sum of the totality of the subjective signs /S(s/), the objective signs /S(o)/ and the instrumental ones /S(i) where each domain of signs can take values from 1 to n. Thus Cf/S1/ is:

$$\sum_{k=1}^{n}(S_k(s) + S_k(i) + S_k(o)) \quad (2)$$

Then, Cf/S1/ faces the options O1, O2, O3, ... On, from which the first decision DM1 is made (**Figure 5**).

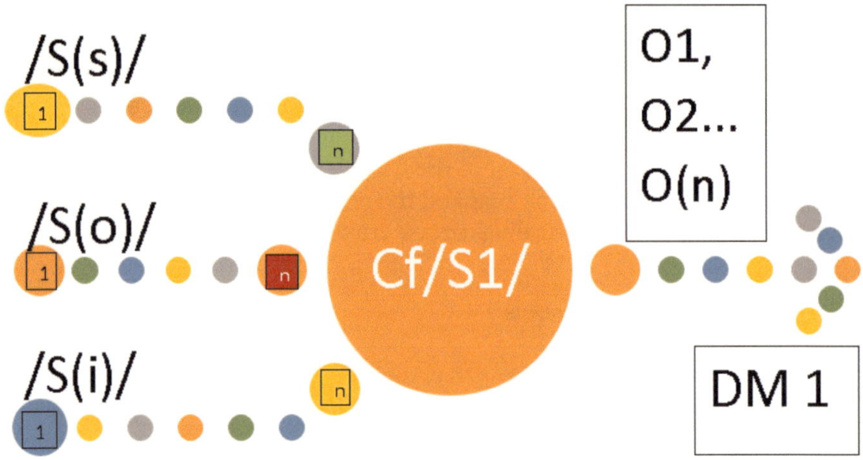

Figure 5.
Subjective, objective, and instrumental signs (S) make signs configurations (Cf) that generate a decision making (DM) process among different options (O).

The first decision leads to the continuation of the process on the next level, for example, from addressability to the medical consultation, where the same structure, with the three types of signs, will lead to a new configuration Cf/S2/, which faces other options, that leads to a new DM2 decision, the one of the diagnosis. Then, the DM3 therapeutic decision will follow that of the DM4 evaluation and the process can continue until the desired result is obtained.

It is worth noting that in a configuration of signs, the signs relate: (a) on the one hand to their *referent*; (b) they relate *to each other* and this valorizes them through association; (c) on the other hand, their value is modulated from *the DM perspective* that will follow (e.g., the perspective of a gout diagnosis makes the relative value of the uric acid analysis potentiate the objective sign /join swelling /). They do not simply add up, but their summation increases the possibility of confronting the normative configuration of the /diagnosis/; in other words, they tend to constitute *meaningful configurations for the DM*, in which purely semiotic interrelationships and those that foreshadow the DM are specified step by step.

What should be emphasized in the case of the former is that they take place in from the first signs that appear in the personal life of a person, before the person becomes a patient, until treatment and then evaluation. Medical decisions are not always simple, but they are always based on configurations of signs that are enriched at each level. If we call the household decision H (home decision), that of Addressability = A, Diagnosis = D, Treatment = T, Evaluation of evolution = E, Follow-up = F, we obtain a course, where each decision is preceded by the formation of a semiotic configuration Cf/S/ (**Figure 6**).

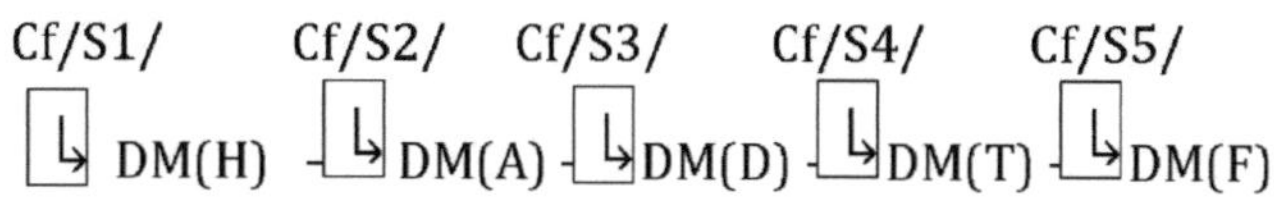

Figure 6.
Successive configurations of signs Cf/S/ lead to successive decision-making DM, from home decision to medical follow up.

After each configuration of signs, a DM follows, based on certain criteria C, and taking into account the options O. Passing from one level of DM to another implies a new configuration of signs, some continuing those from the previous level, others being reconsidered, and others resulting from the anamnesis, observations, or the prescription of paraclinical and laboratory analyses and investigations. For example, the first configuration of signs /S1/ related by patient is not enough for the physician to evaluate the patient and make a DM. They will reevaluate and deepen their evaluation by asking questions about the subjective signs S(s) that thus will become better defined; they will observe the objective signs S(o) and will discover new ones or qualify them. Eventually, they will request and then interpret the required analyses.

It should be noted that the request for analyses and investigations (imaging, genetics, graphic tests, EKG, or EEG) is itself a DM step on the part of the physician, who must decide between several alternatives O: (a) if they must be done; (b) what investigations should be done, so as to bring information that is relevant to the case (e.g., to ask for an HLA B27 test in some persistent joint pains, but not other genetic tests); (c) evaluate the cost-benefit ratio of certain tests (e.g., decide that an expensive MNR is not needed in a panic attack). It is only after the new configuration of signs has been made, with the additions related to the physician's expertise, that the decision regarding the DM(D) diagnosis can be made, and even for this a proper arrangement of signs as criteria can lead to an AHP. But even in this instance, there are one or more intermediate steps, namely those of possible *decisions on the DM(Dd) differential diagnoses*, and the investigations can be done in several stages, as the logic of the diagnosis advances [30, 31].

The final goal of the entire situation that is created in the physician-patient relationship is the treatment. Starting from the patient's signs and taking into account the normative signs from science and the physician's experience, a new problematic configuration is formulated, wherein therapeutic options must be chosen based on certain criteria. Between the medical signs resulting from the consultation and analysis, the diagnostic decision, and the therapeutic decision, there is a connection that is sometimes expressed in models recommended by the respective specialties.

Medical treatments are probably the thorniest problem of medicine and lend themselves best to the AHP because in the very way they are formulated there is a hierarchy of criteria. In all these processes, however, an AHP-type initiated algorithm can be applied, as in the scheme below, yet wherein the number of criteria and options is different from case to case. In essence, it is all about establishing the objective (T = task), a number of criteria (C), from C1 to Cn, and choosing between a number of possible options (O) from O1 to On, then passing through an evaluation of the criteria in the given concrete situation (**Figure 7**).

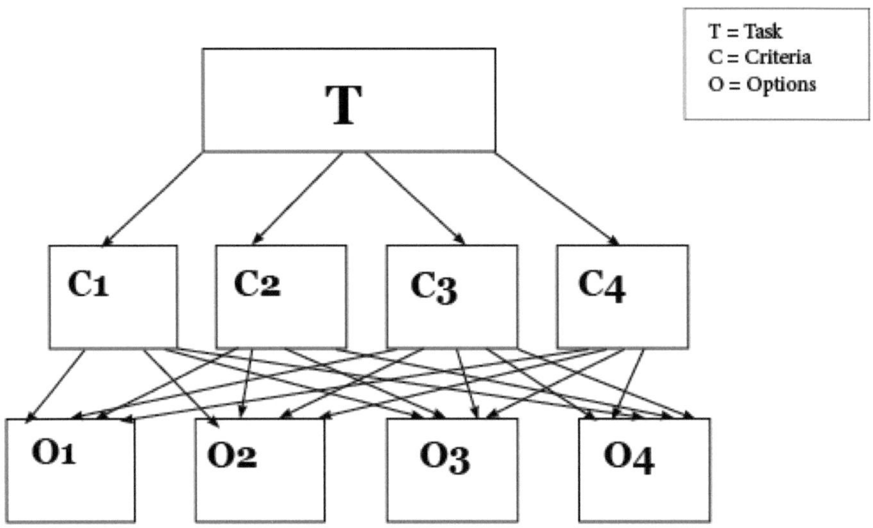

Figure 7.
Hierarchical structure for the AHP modeling of a problem with four criteria (C) and four options (O) the number of criteria and options may vary according to the specific task of DM.

7. Areas in which APH can be applied in medical DM

AHP is an attracting model for DM, and after Saaty published his initial works about AHP but especially after 1997, the model was tested to formalize medical decisions. In a review published in 2008, Liberatore and Nydick considered already 50 articles about AHP classified in seven categories: diagnosis, patient participation, therapy/treatment, organ transplantation, project and technology evaluation and selection, human resource planning, and health care evaluation and policy [32]. For a review of the wider use of AHP, see also Ho [33]. Each stage of an approach in a medical context implicitly or explicitly has a DM correlative, and a possible use of the AHP model. The fields in which they can be used are:

1. Domestic (as in the examples above) and patients medical decisions [34]

2. Addressability to physician or medical institutions, screenings [35].

3. The diagnosis. Any diagnosis is a decision, not a simple linear path or summation of symptoms and signs

4. Treatment with multiple variants, depending on the criteria present in the patient, for example, breast cancer [34] or antibiotic therapy [36].

5. Medical evaluation (follow-up), wherein the treatment and its possible continuations are assessed (e.g., the decision to continue or stop chemotherapy).

6. Quality evaluation of different aspect of medical activity: medical records [37], health supply [38, 39]

7. Public health policies [40], vaccinations [41, 42], health planning policies [43]

The number of medical studies in different fields multiplies fast, as the method is appealing and necessary in complicated decision processes. In a recent (2022) scoping review about AHP use in medicine, the authors conclusion (and we join it) is: "*Despite the compelling rationale on the potential for decision analysis to support shared decision-making, rigorous randomized controlled trials are needed to confirm these interventions' effectiveness, while qualitative studies should seek to understand their potential implementation*" [44].

A general methodological flowchart of some main interactions among Semiotics, DM and AHP is shown in the **Figure 8**.

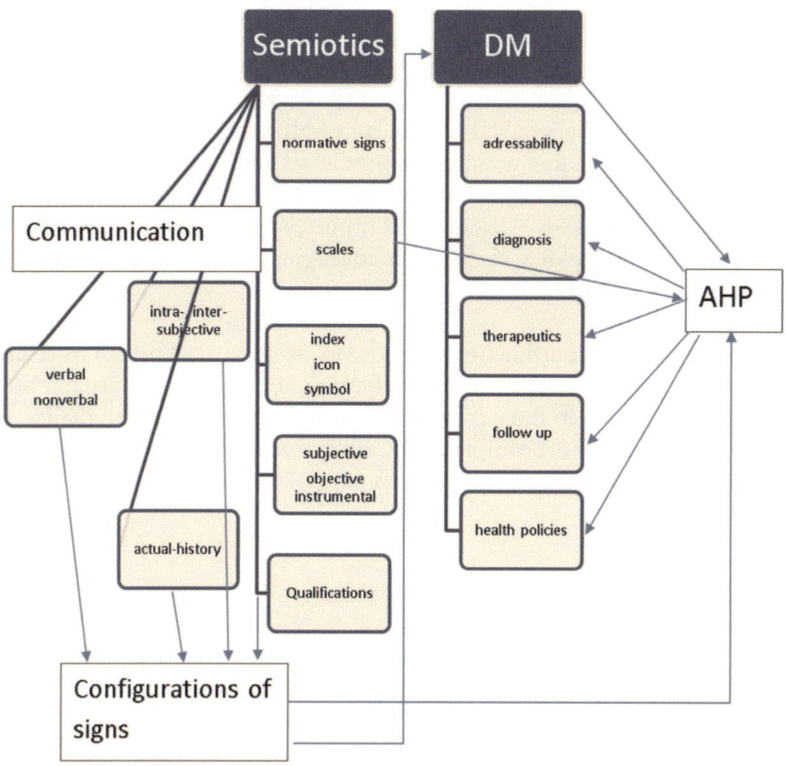

Figure 8.
Semiotics, decision-making, and AHP.

8. Conclusions

The two processes, the process of signifying, of giving meaning, and the one related to making decisions, are permanently inextricably intertwined in medicine:

1. From the meaning given to signs to the domestic decision, the way a person or those around interpret and act in the state of non-well.

2. For addressability to a medical institution, the decision based on the interpretation of certain signs that decide addressability.

3. In diagnosis, each step towards a diagnosis requires mental operations that help to make certain decisions, while the diagnosis itself is a decision.

4. For treatment: from the multitude of treatments, given the individual signs of each patient, the decision of one treatment option or another is mandatory.

5. In the follow-up and management of any medical case and depending on the reaction of the patient or the disease, or other criteria (e.g., cost-benefit), the way of evaluation, continuation, interruption of treatment is decided (interruption of treatment is a clear case of DM, based on semiotic criteria).

6. To state or regional medical policies, for example, vaccination, anti-smoking campaigns, or reduction of schooling, corporate stress, etc.

7. To the attitudes and the construction of a certain medical paradigm within a community, either at the state level or at the level of certain micro-communities that share the same beliefs and values. By interpreting certain social signs in a certain way, attitudes and actions are predicted and campaigns are conducted (e.g., "drink water regularly," "excessive alcohol consumption damages your health," "get vaccinated," "exercise 30 minutes a day," etc.) and promote a certain medical image of society, by encouraging values or denigrating habits, beliefs, and activities.

8. It finally leads to the outline of one of the most important health problems, that of the /normal/ as a reference image whereby normality is described, not only in the medical, but also in the paramedical problems (e.g., homosexuality, now treated as becoming part of normality, after it was previously considered a pathology or abnormality, compared to natural norms).

Acknowledgements

I express my gratitude to Oana Georgiana Gabor who helped me in the translation and arrangement of the English text, and to Bogdan Carasca for his help in mathematical processing of the example of AHP as domestic medical DM.

Author details

Gheorghe Jurj[1,2,3]

1 Medical Univ Timisoara, Timișoara, Romania

2 West Univ Timișoara, Timișoara, Romania

3 Asociatia Romana de Homeopatie Clinica, Romania

*Address all correspondence to: relujurj@gmail.com

References

[1] Von Uexkull T. Semiotics and medicine. Semiotica. 1982;**38**(3–4): 205-216. DOI: 10.1515/semi.1982.38.3-4.205

[2] Von Uexkull T. Medicine and semiotics. Semiotica. 1986;**61**(3–4): 201-218. DOI: 10.1515/semi.1986.61.3-4.201

[3] Hoopes J. In: Hoopes J, editor. Peirce on Signs, Writing on Semiotics by Charles Sanders Peirce. The University of North Carolina Press. 1991

[4] Morris CW. Writings on the General Theory of Signs, (Approaches to Semiotics [As]). De Gruyter Mouton

[5] Eco U. A Theory of Semiotics. Indiana University Press. 1976. Available from: http://www.jstor.org/stable/j.ctt16xwcfd

[6] Sebeok TA. The Sign & Its Masters. University Press of America. 1989

[7] Jakobson R. Essais de Linguistique générale. Les Edition de Minuit. 1963

[8] World Health Organization. Available from: https://www.who.int/about/governance/constitution

[9] Canguilhem G. Le normal et le pathologique. Presses Universitaires de France. 2013. DOI: 10.3917/puf.cangu.2013.01

[10] Foucault M. El Nascimento de la Clínica; Siglo xxi editores, 17a edición esp. 1997

[11] Catita M, Águas A. & Morgado P: Normality in medicine: A critical review. Philosophy, Ethics, and Humanities in Medicine. 2020;**15**:3. DOI: 10.1186/s13010-020-00087-2

[12] Hofmann B. Normalitet og patologi [The normal and the pathological]. Tidsskr Nor Laegeforen. 1995;**115**(30): 3800-3802

[13] Tiles M. The Normal and pathological: The concept of a scientific medicine. The British Journal for the Philosophy of Science. 1993;**44**(4): 729-742

[14] Zajicek G. Normative medicine. Medical Hypotheses. 1995;**45**(4): 331-334. DOI: 10.1016/0306-9877(95)90090-X

[15] Ladino S. La Semiótica de Peirce aplicada a la semiología médica: El "Infarto", el "Electrocardiograma" y la "Quinoterapia" como fenómenos de referencia. Miradas. 2014;**1**(12):89-98

[16] Jurj G. Decision making and semiotics: A view in homeopathy. International Journal of High Dilution Research. 2008;7(23):103-112

[17] Saaty TL. A scaling method for priorities in hierarchical structures. Journal of Mathematical Psychology. 1977;**15**(3):234-281

[18] Cirino E. Pain scale. https://www.healthline.com/health/pain-scale

[19] Smith RC, Hope RB. The patient's story: Integrating the patient- and physician centered approaches to interviewing. Annals of Internal Medicine. 1991;**115**:470-477

[20] Jurj G. Signs in the medical context and the semiotic codes: An introduction to medical semiotics. Cultura Homeopatica. 2007;**nr 20**:32-36

[21] Dolan JG, Isselhardt B, Cappuccio JD. The Analytic Hierarchy

Process in Medical Decision Making: A Tutorial. DOI: 10.1177/0272989X8900900108

[22] Dolan JG, Bordley DR. Using the Analytic Hierarchy Process (AHP) to Develop and Disseminate Guidelines, QRB. Quality Review Bulletin. 1992; **18**(12):440-447

[23] Nica S et al. Addressability to the emergency department of patients with acute post-COVID 19 clinical symptoms. Romanian Journal of Military Medicine. 2022;**125**(2):275

[24] Jurj G. C.S. Peirce and the Deconstruction of Tradition, Reality as a Sign. Cambridge Scholars Publishing. 2023

[25] Buzzelli D E : The Argument of Peirce's 'New List of Categories. Transactions of the Charles S. Peirce Society, 8.2: 63–89, 1972. http://www.jstor.org/stable/40320362 [Accessed Jan. 27, 2023]

[26] Jurj G. Semiotic processes implied in phaenomena associated with high dilutions. International Journal of High Dilution Research. 2010;**9**(31):74-84

[27] Moradi K et al. Effective components on the process of concept and meaning creation and their application in ontologies: Focusing on Peirce's triadic sign model. The Electronic Library. 2022; **40**(4):486-502. DOI: 10.1108/EL-10-2021-0185

[28] Nessa J. About signs and symptoms: Can semiotics expand the view of clinical medicine? Theoretical Medicine and Bioethics. 1996;**17**:363-377. DOI: 10.1007/BF00489681

[29] John F. Burnum: Medical diagnosis through semiotics: Giving meaning to the sign. Annals of Internal Medicine. 1993;**119**:939-943. DOI: 10.7326/0003-4819-119-9-199311010-00012

[30] Dolan JG. Can decision analysis adequately represent clinical problems? Journal of Clinical Epidemiology. 1990; **43**:277-284

[31] Dolan JG, Bordley DR. Bedside decision making using the analytic hierarchy process. Medical Decision Making. 1992;**12**:344

[32] Liberatore MJ, Nydick RL. The analytic hierarchy process in medical and health care decision making: A literature review. European Journal of Operational Research. 2008;**189**(1): 194-207

[33] Ho W. Integrated analytic hierarchy process and its applications—A literature review. European Journal of Operational Research. 2008;**186**(1):211-228

[34] Dolan JG, Bordley DR. Individualized patient decision making using the analytic hierarchy process (AHP): Reliability, validity, and clinical usefulness. Medical Decision Making. 1991;**11**:322

[35] Liberatore et al. Decision counseling for men considering prostate cancer screening. Computers & Operations Research. 2003;**30**(10):1421-1434

[36] Mostafa H, Büyüktahtakın I, Elshami E. A :Multi-criteria ranking algorithm (MCRA) for determining breast cancer therapy. Omega. 2019;**82**: 83-101

[37] Dolan JG. Medical decision making using the analytic hierarchy process: Choice of initial antimicrobial therapy for acute pyelonephritis. Medical Decision Making. 1989;**9**(1):51-56. DOI: 10.1177/0272989X8900900109

[38] Ajami S, Ketabi S. Performance evaluation of medical records departments by analytical hierarchy process (AHP) approach in the selected hospitals in Isfahan. Journal of Medical Systems. 2012;**36**:1165-1171. DOI: 10.1007/s10916-010-9578-9

[39] Zamiela C et al. Enablers of resilience in the healthcare supply chain: A case study of U.S healthcare industry during COVID-19 pandemic. Research in Transportation Economics. 2022;**93**: 101174

[40] Shweta and Kumar D. Analysis of issues of generic medicine supply chain using fuzzy AHP: A pilot study of Indian public drug distribution scheme. International Journal of Pharmaceutical and Healthcare Marketing. 2021;**15**(1): 18-42. DOI: 10.1108/IJPHM-12-2019-0078

[41] Byun D et al. Prioritizing community-based intervention programs for improving treatment compliance of patients with chronic diseases: Applying an analytic hierarchy process. International Journal of Environmental Research and Public Health. 2021;**18**:455. DOI: 10.3390/ijerph18020455

[42] Shin T et al. The comparative evaluation of expanded national immunization policies in Korea using an analytic hierarchy process. Vaccine. 2009;**27**(5):792-802

[43] Odynocki B. The forward and backward processes in health policy planning. Mathematics and Computers in Simulation. 1983;**25**(2):146-155

[44] Humphries B et al. More work needed on decision analysis for shared decision-making: A scoping review. Journal of Clinical Epidemiology. 2022; **141**:106-120. DOI: 10.1016/j.jclinepi.2021.10.001

Section 2

Applications and Use Cases

Chapter 7

Multi-Criteria Decision-Making in the Transport Sector

Federico de Andreis, Enza Curcio, Federico Maria Sottoriva and Ubaldo Comite

Abstract

The transportation sector stands within a highly competitive environment, continuously seeking for effectiveness and efficiency. In the logic of decision-making processes, optimization—the selection of the best element, with regard to some criterion, from some set of available alternatives—appears central. In addition to traditional decision making, there is the technique of Linear Programming, which is an optimization problem in which the objective function is a linear function subject to linear constraints, which may be equalities and/or inequalities. Many real-life problems can be modeled in terms of Linear Programming, highlighting its strong practical interest. Therefore, Linear Programming can be a useful tool to support management in decision-making activities, helping both to reduce human effort and to provide quantitative results to optimization problems, such as revenue maximization or cost minimization. The aim of this research is to highlight how the combination of qualitative methods and quantitative techniques is an ideal combination for identifying the best solution in the decision-making process, contributing to research in the fields of organizational behavior and management. The chapter analyzes the whole decision-making process in transportation organizations seeking competitive advantage. Through a practical application of Linear Programming, presented using a case study, the chapter results demonstrate the importance of an instrument supporting decision-making process.

Keywords: decision-making, transport industry, multi-criteria decision-making, linear programming, analytic hierarchy process, organization

1. Introduction

The transport sector is of great importance for the economic development of a country, both with regard to the transport of people and goods [1–3]. It follows that the ability to easily offer services also increases the productivity of companies.

There are many innovative elements brought about by the rapid change increasingly evident to society, but also elements that could be improved.

The future prosperity of the industry will depend on the ability of all companies to remain competitive within the global economy [4]. It is therefore necessary to remain efficient by examining problems that may arise and, in turn, seeking innovative and optimal solutions [5].

In response to this need, organizations have to make decisions and it is precisely in organizations that decision-making processes are a very difficult aspect to evaluate.

Both in the case of individual decision-makers and in the case of groups, in fact, making the correct decision is very complex.

The definition of the problem, the alternatives, the method of evaluation, and the limitations and errors of the decision-makers, push organizations toward decisions that are not always optimal [6].

With this purpose, using multiple criteria decision-making appears to be one of the best solution for organizations in order to be competitive and to take the best decision, facing a continuously changing environment, such as in the transport industry [7].

Multiple criteria decision analysis is a structured process for evaluating options with conflicting criteria and choosing the best solution in a similar way to a cost-benefit analysis but evaluating also numerous criteria, rather than just cost [8].

Furthermore, conducting an MCDA aims to help the organization determine which options are most effective, increasing the efficiency of the decision-making process [9]. In addition to provide an ordered list of alternatives, it addresses the social aspects of decision-making to encourage discussion between different decision-makers.

The chapter, therefore, aims to analyze how more scientific models of analysis can complement the organizational decision-making approach, allowing decisions to be made that are more correct for creating value and gaining an advantage over competitors.

The paper is structured as follows. In Section 2, the methodology of the research review is set out. Next, the findings of the literature analysis on the decisions in the organizations and the related decision-making models are shown in Sections 3 and 4. In Section 5, Linear Programming is described. In Section 6, the discussion and the case study, drawn from this method, are presented. In Section 7, the results of the research are shown. Finally, conclusions and implications for future research are presented in Section 8.

2. Methodology

The methodology used in the paper approaches a literature review analysis on decision-making in complex organizations, such as aviation.

Following the description of qualitative methods of decisions, quantitative methods, such as Linear Programming—LP and Analytic Hierarchy Process—AHP are presented.

To verify the hypothesis a case study is presented [10]. The case study demonstrates that the addition of quantitative methods to qualitative ones can be considered the best solution in order to take proper decisions in organizations.

3. The decisions in the organizations

Organizational life is marked by a series of decisions taken at different hierarchical levels. For example, the relational climate, short- and long-term strategies, process trends, and the quality of working life [11].

The main characteristics of an organization constitute the decision dimensions. The latter influence each other and identify both the nature and the type of the decision itself.

The three main dimensions of a decision are relevance, temporality, and context:

Relevance: It is important to specify that organizational decisions can be simple routine matters but also changes in the whole strategy. Therefore, it can be deduced that the relevance of a decision can specify the impact on the whole organization.

Temporality: It expresses the period of time in which its consequences will be felt, given that a decision can have an immediate or distant effect in time.

Context: The environmental conditions in which a decision is made can vary significantly, influencing the possibility of finding the information necessary to define the problem and possible solutions.

In this regard, it is important to specify that there are different types of situations: certainty, risk, and uncertainty. In situations of certainty, it is possible to foresee the outcome of the decision in advance, as there is total knowledge of the information; in risk situations, it is only possible to make projections on the possible outcome of the decision, as there is partial knowledge of the information; in situations of uncertainty, it is not possible to project the possible outcome of the decision, as there is not enough information.

Starting from the interaction of these three dimensions, the economist Herbert Alexander Simon has identified two large families of decisions: planned decisions and unplanned decisions [12].

On the one hand, the planned decisions are repetitive and routine, addressing structured, well-known, and familiar problems. In fact, it is possible to elaborate a defined procedure, related to problems that occur with a certain repetitiveness and frequency. Furthermore, planned decisions are also defined as operational decisions because they have short-term effects and therefore can be taken at all hierarchical levels of the organization.

On the other hand, unplanned decisions deal with unstructured problems, that is, unexpected situations about which very little information is available. Consequently, it is not possible to manage them by adopting standard procedures: they require an authentic, original, and innovative solution. Unscheduled decisions deal with matters of great importance to the organization. Unplanned decisions are divided into tactical decisions and strategic decisions: the former address issues with short- and medium-term effects that require a dose of creativity and improvisation; the latter have broad relevance and a high level of risk because they modify long-term strategies, in fact, they require all the attention and creativity available to the organization.

4. Decision-making models

Every day and several times a day each of us makes decisions, and this gives us the possibility to choose between several alternatives: this process is called decision-making. It is a complex process because it involves different cognitive structures in which the individual must evaluate and interpret events in order to choose with greater awareness.

When we have to make a decision, we consider and integrate a lot of information to identify the most appropriate thinking strategies to decide on and to generate choice alternatives [13]. Therefore, deciding means arriving at a definitive judgment after having weighed and taken into consideration a series of alternatives and possible choices.

Decision-making is a reasoning process that can be done in a planned or unplanned way.

The study of decision-making dates back to the early 1950s: the main purpose of the research was to describe how a person should make decisions by behaving rationally. This prescriptive approach, defined as normative or rational, provides for the optimization of available resources and refers to the complete rationality of the decision-making process.

The effectiveness of rational decision-making is based on several assumptions, such as the absolute rationality of the decision-maker, the irrelevance of his emotional state, the availability of resources, and the independence of the decision-maker from the environment in which he operates in order to be able to evaluate more information.

Normative decision models are based on the notion of expected value, which consists of what the decision-maker can expect from each choice option. According to this principle, the rationality of the decision-maker is evaluated on the basis of the maximization of a monetary value intended as an advantage of the choice made.

The principle of expected value has proved to be inadequate as it is not always possible to convert a result into monetary value and the latter can have a different value for different people.

Bernoulli highlighted a concept called "moral value," according to which decisions are determined by the utility that the outcomes have for the decision-maker and not only by their monetary value [14].

The weakness of regulatory models consists in not considering the limits of the human decision-maker and in not considering the decision-making context and the limited capacity of the cognitive system of the decision-maker in processing information.

On the other hand, since the early 1970s, decision psychology has moved toward a descriptive approach, which produced the first results with a work conducted by Herbert Alexander Simon in 1967 in which the effectiveness of the rationality model was demonstrated limited: Simon represented the decision-maker as an infallible scientist who has limited and intentional rationality. Indeed, according to Simon, individuals are limited by internal and external constraints which can be traced at different levels.

Furthermore, also Lichtenstein and Slovic deepened the so-called phenomenon of the reversal of preferences. Thanks to the deepening of this phenomenon, it has been possible to demonstrate and observe in some experiments that, when subjects have to choose between two bets, they prefer the one with the highest probability of winning, which however guarantees a small win. However, when asked to indicate how much they would pay for these same bets, they set a higher price for the bet with the lowest probability of winning, but which offers a higher payout [15, 16].

Over the years we have witnessed the birth of new theoretical orientations that seek to develop models and theories no longer conditioned by the need to check the validity of the principles underlying the rational and normative behavior of the decision-maker. In fact, starting from prospect theory, the tendency to explain decision-making behavior on the basis of the representation of the decision-making context has been emerging. The latter depends on a series of interconnected factors, such as social, moral, and individual values. It can therefore be deduced that the fundamental nature of the decision-making task is no longer the "choice" between the available alternatives based on the value of their expected utility, but the construction of the reasons for a choice in relation to another possible choice [17].

The centrality of decision-making in the competitive advantage that organizations can achieve certainly includes choices but also the decision not to choose. Such decisions, however, cannot simply refer to the "feeling" of management, although certainly "the idea" remains the constituent engine of the enterprise, but must be supported by increasingly analytical models.

Decision-making processes, at all levels, are incredibly complex and—since not everyone can have a visionary capacity—intelligence in managing information through data analysis becomes a fundamental competitive advantage for all those who want to make a complex organization such as that of a large company function properly.

Indeed, in a market that can hardly be understood in a general overview, given its complexity, the goal of those with real decision-making power is to obtain observations based on data analysis.

Now that various tools have made storage decidedly cheaper and make in-depth analyses of data possible in real time, the use of insights obtained from large samples can provide evidence where even a very good observer would fail to arrive. Therefore, Operation Research could complement normative models.

Operations Research, also known as Decision Theory, is that branch of applied mathematics in which complex decision-making problems, which arise in the management of organizations and enterprises, are analyzed and solved by mathematical models and advanced quantitative methods (optimization, simulation, etc.).

The objective of Operations Research is to support decision-making.

To achieve this goal, Operations Research provides mathematical tools to support decision-making activities in which limited resource activities must be managed and coordinated in order to maximize or minimize an objective function.

5. Linear programming

As previously stated, the purpose of organizations is the creation of value in order to gain a competitive advantage in the reference market.

The affirmation of the market, the stability, and the growth of an organization are closely linked to the critical and strategic decisions of management, such as the choice of prices, the identification of the target market, the specialization in the core activities and the allocation of the necessary resources to carry out the activities efficiently. Management will therefore frequently face optimization problems, defined as the selection of the best element, with regard to some criterion, from a set of available alternatives.

Therefore, the result of the optimization problem is the optimum solution, which represents the best, or most favorable condition, possible under specific sets of comparable circumstances.

Linear Programming (LP) is a subset of mathematical programming that aims to efficiently allocate limited resources to known activities to achieve a desired profit maximization or cost minimization. In statistics and mathematics, Linear Programming is a method for optimizing a linear objective function that may be subject to both linear equality and inequality and constraints.

Practically, Linear Programming represents an instrument to achieve the best outcome in a given mathematical model, composed of a list of requirements described as linear equations and a set of linear constraints [18].

The application of Linear Programming has a very wide domain, from business/economics to engineering problems. Our focus will be on its operational utility in modeling problems concerning planning, routing, and scheduling assignment in strategic operations contexts. To make a practical example, Linear Programming makes it possible to answer the following frequent questions for production management:

- What is the maximal production size?
- What is the structure of the production?
- What are the optimal production size and structure?
- What will be the profit reduction in case of the shortcoming of some units of material?
- Which production material should be bought in order to reach the maximum increase in profit?
- What are the possibilities for price negotiation?

5.1 Linear programming problem

The standard Linear Programming model consists of three main elements:

- Objective function (Eq. (1)), which is the function we want to maximize (profit) or minimize (cost).

$$\text{Min } f(x) = c_1x_1 + c_2x_2 + \ldots + c_nx_n. \tag{1}$$

c = cost coefficient of the objective function, x = decision variables

- Decision variables (x)
- Constraints (v), conditions that must be satisfied when optimizing the objective function, expressed in algebraic equations (Eq. (2)) (equalities or inequalities).

An important assumption to consider is that activities cannot be negative (x > =0).

$$v_i = \alpha_1x_1 + \ldots \alpha_nx_n > = b_i, i = 1, \ldots., m. \tag{2}$$

α = coefficient, b = known terms of constraints.

To formalize the problem in a mathematical form a linear system (Eq. (3)) is constructed, considering n decision variables (x), n cost coefficients (c), m constraints (v), and m known terms (b):

$$\begin{cases} min\ f(x) = c_1x_1 + c_2x_2 + \ ...\ + c_nx_n. \\ \alpha_1, 1x_1 + \ ...\ + \alpha_1,_nx_n > \ = b_1 \\ \vdots \\ \alpha_{m,1}x_{m,1} + \ ...\ + \alpha_{m,n}x_n > \ = b_m. \end{cases} \tag{3}$$

In the following section, a practical demonstration of how to optimize operations in an airline company will be presented. The aim of this example will be to solve the following problem: optimize the route scheduling for the budget airline operating between five main airports in the US, Atlanta, Los Angeles, Chicago, Dallas, and New York.

6. Discussion

6.1 Case study: airline operations optimization using linear programming

In order to demonstrate what has been presented previously, the following case study deals with optimization of airlines route scheduling, specifically, looking for the best possible solution in order to maximize profits, considering the routes' fixed and variable costs and profits and some constraints related to the need for providing a connection service between the cities.

Table 1 contains information about the distance between the cities.

Cities	Atlanta	Los Angeles	Chicago	Dallas	NY
Atlanta	—	1942	607	730	759
Los Angeles	1942	—	1741	1232	2469
Chicago	607	1741	—	802	738
Dallas	730	1232	802	—	1389
NY	759	2469	738	1389	—

Table 1.
Distance between the cities.

Table 2 shows the total fuel (in US gallons) required to fly each leg.

Cities	Atlanta	Los Angeles	Chicago	Dallas	NY
Atlanta	—	4097	1623	1851	1906
Los Angeles	4097	—	3724	2782	5073
Chicago	1623	3724	—	1985	1867
Dallas	1851	2782	1985	—	3072
NY	1906	5073	1867	3072	—

Table 2.
Total fuel required for leg (in US gallons).

The cost of fuel is 4.5 USD/gallon. We also consider other fixed costs, such as staff, for 3000 USD for each flight.

Table 3 contains the fare charged for a one-way flight between the cities for one passenger.

Cities	Atlanta	Los Angeles	Chicago	Dallas	NY
Atlanta	—	219$	85$	119$	163$
Los Angeles	219$	—	115$	145$	262$
Chicago	85$	115$	—	125$	105$
Dallas	119$	145$	125$	—	100$
NY	163$	262$	105$	154$	—

Table 3.
Fares for the route.

To simplify the example, we will use the same number of passengers for each flight. Considering the 85% of a common liner jet capacity seat taken, each flight will have 160 passengers.

Another important factor to consider is the number of slots available. Slots are the rights granted by the airport owner to schedule a flight to or from the airport at a specific time. **Table 4** shows the number of slots available in our case study.

	Max slots for departure	Max slots for arrivals
Atlanta	18	14
Los Angeles	25	25
Chicago	12	16
Dallas	14	15
NY	17	16

Table 4.
N° of slots available.

Finally, since the airline must provide a public service, we must ensure that at least one flight is granted between the cities.

Our Linear Programming problem may now be defined as (**Table 5**):

- Aim: Maximize profits (find the best flight schedule which maximizes profits)
- Profits = Total income – Total expenditure
- Total income = Fare per person x N° of seats
- Total expenditure = Fuel required x Cost of fuel + Fixed costs.

Subject to 30 constraints:

- Number of slots for departure (x 5)

- Number of slots for arrival (x5)
- Each decision variable must be > = 1 (x20), to ensure that at least one flight will connect to each city.

Decision variables

	Atlanta	Los Angeles	Chicago	Dallas	NY
Atlanta	0	X12	X13	X14	X15
Los Angeles	X21	0	X23	X24	X25
Chicago	X31	X32	0	X34	X35
Dallas	X41	X42	X43	0	X45
NY	X51	X52	X53	X54	0

Profit coefficients (USD)

	Atlanta	Los Angeles	Chicago	Dallas	NY
Atlanta	0	13,603.5	3296.5	7710.5	14,503
Los Angeles	13,603.5	0	−1358	7681	16,091.5
Chicago	3296.5	−1358	0	8067.5	5398.5
Dallas	7710.5	7681	8067.5	0	−824
NY	14,503	16,091.5	5398.5	7816	0

At least 1 flight per leg constraint

	Atlanta	Los Angeles	Chicago	Dallas	NY
Atlanta	0	1	1	1	1
LA	1	0	1	1	1
Chicago	1	1	0	1	1
Dallas	1	1	1	0	1
NY	1	1	1	1	0

Slots constraint

DV	Atlanta	Los Angeles	Chicago	Dallas	NY	Total departures
Atlanta						< = 18
Los Angeles						< = 25
Chicago						< = 12
Dallas						< = 14
NY						< = 17
Total arrivals	< = 14	< = 25	< = 16	< = 15	< = 16	

Table 5.
Linear Programming problem data.

Through a Linear Programming solver tool, it is possible to easily find the optimum solution in order to obtain the maximum possible profit, which in our case study represents the best flight schedule under the abovementioned conditions (**Table 6**).

Optimum solution: N° of flight for each route								
DV	Atlanta	Los Angeles	Chicago	Dallas	NY	Total departure		MAX DEP
Atlanta	0	9	3	4	2	18	<=	18
Los Angeles	11	0	1	1	12	25	<=	25
Chicago	1	1	0	9	1	12	<=	12
Dallas	1	1	11	0	1	14	<=	14
NY	1	14	1	1	0	17	<=	17
Total arrivals	14	25	16	15	16			
	<=	<=	<=	<=	<=			
Max ARR	14	25	16	15	16			

Table 6.
Optimum solution.

Then, the Objective Function will be: Maximize total profit (leg profit coefficient x N° of flights) = 977.482 USD.

6.2 Sensitivity analysis

Based on the results obtained from the resolution of the Linear Programming problem, it is possible to carry out an analysis of the data to verify the sensitivity of the optimal solution to any changes.

This operation is called Sensitivity Analysis. Through this analysis it is possible to obtain important information in practice, allowing the decision maker to know the degrees of freedom of the problem and to answer "What if" type questions [19] such as: if the objective function changes, how does the solution change? If available resources change, how does the solution change?

In particular, we can focus on two main aspects: the analysis of how the variation of the decision variables influences the optimal solution and the impact of the constraints on the objective function.

First, we will analyze the variables sensitivity (**Table 7**).

Name	Final value	Reduced cost	Objective coefficient	Allowable increase	Allowable decrease
Atlanta Atlanta	0	−12,015	0	12,015	INF
Atlanta Los Angeles	9	0	13,603.5	0	10,307
Atlanta Chicago	3	0	3296.5	4414	386
Atlanta Dallas	4	0	7710.5	3653.5	1618
Atlanta NY	2	0	14,503	1618	0
Los Angeles Atlanta	11	0	13,603.5	INF	0
Los Angeles Los Angeles	0	−15,192	0	15,192	INF
Los Angeles Chicago	1	−6243	−1358	6243	INF
Los Angeles Dallas	1	−1618	7681	1618	INF

Name	Final value	Reduced cost	Objective coefficient	Allowable increase	Allowable decrease
Los Angeles NY	12	0	16,091.5	0	1618
Chicago Atlanta	1	−9075.5	3296.5	9075.5	INF
Chicago Los Angeles	1	−15,318.5	−1358	15,318.5	INF
Chicago Chicago	0	−3653.5	0	3653.5	INF
Chicago Dallas	9	0	8067.5	INF	3653.5
Chicago NY	1	−9461.5	5398.5	9461.5	INF
Dallas Atlanta	1	−9075.5	7710.5	9075.5	INF
Dallas Los Angeles	1	−10,693.5	7681	10,693.5	INF
Dallas Chicago	11	0	8067.5	INF	8067.5
Dallas Dallas	0	−12,481.5	0	12,481.5	INF
Dallas NY	1	−20,098	−824	20,098	INF
NY Atlanta	1	0	14,503	0	INF
NY Los Angeles	14	0	16,091.5	INF	0
NY Chicago	1	−386	5398.5	386	INF
NY Dallas	1	−2382.5	7816	2382.5	INF
NY NY	0	−16,991	0	16,991	INF

Table 7.
Sensitivity report: variables analysis.

The final value represents the optimal resource allocation solution. This is the number that will replace the decision variable X in the objective function.

Reduced cost is the decrease of the objective function value for the production of one unit of the product. Reduced cost is a negative amount for maximization problems and represents the loss in profits if one unit of the activity is forced into the solution.

There might be situations where not producing one product (activity) is more profitable. In this case, the final value will be zero.

In our case study, we can identify routes where reduced cost is a negative number, which means that removing that flight from the schedule will improve the total profit. However, this decision is incompatible with the "at least one flight per city" constraint and therefore cannot be taken into account.

The allowable increase or decrease shows the range between the objective coefficient can change so that the optimum solution remains unchanged. Referring to the case study, the objective coefficient is the profit per route, which can fluctuate between an upper and a lower limit indicated by the allowable increase/decrease number.

If the objective coefficient varies outside the allowable interval, there will be a new optimum solution, different from the original one.

Now we can move to the constraints analysis (**Table 8**).

The shadow price, also known as opportunity cost, is the amount by which the value of the objective function will improve or decrease if the availability of the resources associated with the related RHS (right-hand side) constraint is increased or reduced by a unit.

Name	Final value	Shadow price	Constraint R.H. side	Allowable increase	Allowable decrease
Atlanta	14	8718.5	14	2	0
Los Angles	25	10,307	25	2	0
Chicago	16	0	16	INF	0
Dallas	15	4414	15	2	0
NY	16	11,206.5	16	2	0
Atlanta	18	3296.5	18	0	2
Los Angeles	25	4885	25	0	2
Chicago	12	3653.5	12	0	2
Dallas	14	8067.5	14	0	10
NY	17	5784.5	17	0	2

Table 8.
Sensitivity report: constraints analysis.

This information is very useful because it allows decision makers to know the effective impact on the profit of a change in terms of constraints just by multiplying the shadow price by the increase or decrease quantity of the RHS constraint, without having to solve a new Linear Programming problem.

Also, in this case, there is a range of feasibility, identified by the allowable increase and decrease value, within which the RHS constraint may vary, maintaining the shadow price unchanged. If the variation of the RHS constraint falls outside the upper or lower limit, the shadow price will no longer be reliable, making it impossible to evaluate the impact on the profit of that change unless we build up a new Linear Programming problem.

The sensitivity analysis, therefore, reveals to be an essential tool for the post-optimal evaluation of possible new scenarios. For example, in our case study, the strategic need to buy or sell a slot could be a plausible scenario.

Through this analysis, it is quite easy to take the best decision that allows the company to maximize the increase in profits or minimize losses in case of sale.

6.3 Analytic hierarchy process vs. linear programming

In high complexity and uncertainty contexts, such as the transport industry, especially when facing different alternatives that must meet multiple objectives and different stakeholders' needs, it is necessary to rely on analysis methods that support the decision-making process in order to take the best decisions.

Depending on the context, the needs and the objectives to be achieved, it is essential to choose the most appropriate method to support decision-making.

The Analytic Hierarchy Process (AHP) is a multi-criteria decision analysis methodology capable of suggesting the best solution among a discrete set of alternatives, using mathematics and psychology. The method attributes priority values to individual alternatives, determined through a multilevel hierarchical structure that weighs their importance through quantitative and qualitative judgments.

The structure of the Analytic Hierarchy Process is developed on three levels:

- The final goal to be achieved (the problem to be solved);
- The set of possible solutions called alternatives;
- The decision-making criteria through which the goodness of the alternatives to achieve the goal is evaluated.

The choice of criteria and the judgments used to evaluate the characteristics of the alternatives are subjective interpretations, often of a qualitative nature and expressed in verbal language, which is subsequently transformed into numerical values using the Saaty ratio scale [20].

This conversion table (**Table 9**) assigns a value from 1 to 9 to the evaluation criteria, where 1 represents the equality between the criteria and 9 the clear superiority of one over the other. The final value is calculated using the weighted average of all judgments.

Intensity of importance	Definition	Explanation
1	Equal importance	Two activities contribute equally to the objective.
3	Weak importance of one over another	Experience and judgment slightly favor one activity over another.
5	Essential or strong importance	Experience and judgment strongly favor one activity over another.
7	Demonstrated importance	An activity is strongly favored and its dominance is demonstrated in practice.
9	Absolute importance	The evidence favoring one activity over another is of the highest possible order of affirmation.
2, 4, 6, 8	Intermediate values between the two adjacent judgments	When compromise is needed.

Table 9.
Saaty's scale of pairwise comparison (compiled by the author. Source: ([20], p. 246)).

All the decision-making elements belonging to the same hierarchical level are then compared in pairs, thus obtaining a preference ratio that will highlight the local priorities. Considering all the local priorities of the single levels, the principle of hierarchical composition is applied [20] to calculate the priorities of the alternatives, called global priorities.

The Analytic Hierarchy Process is therefore structured in a series of steps that allow the initial problem, complex and unstructured, to be broken down into a rational framework that is simpler to understand and evaluate.

However, this method expects people, as decision-makers or stakeholders, to express preferences. Therefore, the judgments will vary from person to person and the final result will also be influenced by the subjectivity in the preferences of those who participated in the assignment of judgments.

Summarizing, we can state that the Analytic Hierarchy Process helps decision-makers to identify the alternative that best suits their values and their understanding of the problem.

Unlike the Analytic Hierarchy Process, Linear Programming provides an objective optimal solution, calculated exclusively on the final goal and on the conditions imposed on the system, which is not influenced by the subjectivity of the decision maker.

However, this characteristic is not necessarily positive; depending on the needs of the individual cases, it may sometimes be necessary to insert a preference criterion for selecting the various alternatives, thus making the Analytic Hierarchy Process preferable to Linear Programming.

To obtain more precise results and better adapt the quantitative decision support model to the problem, it is possible to integrate the two models, so obtaining an integrated Analytic Hierarchy Process-Linear Programming model (AHP-LP), which allows to insert a degree of flexibility in the search for the optimal solution, represented by the subjective weights attributed in the Analytic Hierarchy Process.

Analytic Hierarchy Process is capable of evaluating both quantitative and qualitative factors, and is used to build a priority scale according to the subjective judgments of decision-makers. Afterward, the priority ratings obtained from the Analytic Hierarchy Process model are used as coefficients of the decision variables in a Linear Programming model to find an optimal solution that maximizes the desired benefits [21].

7. Results

The study demonstrated how qualitative and quantitative methodologies work well together to find the optimal option throughout the decision-making process. The methodology employed in the study concerns the research on decision-making in complex organizations, such as the aviation industry.

After describing qualitative decision-making techniques, quantitative techniques, such as Linear Programming (LP) and the Analytic Hierarchy Process (AHP), are introduced.

Then, the paper analyzed the entire decision-making process in transport companies in search of competitive advantage, the study results illustrate on one hand the importance of a quantitative instrument to support the decision-making process; on the other hand, how the latter helps to identify some choices that turned out to be better than others, through a practical application of Linear Programming.

Linear Programming, as proved in the case study, is able to effectively distribute precious resources to well-known activities in order to maximize profits or minimize costs.

The case study "Airline operations optimization using Linear Programming" has been presented to support the theory: this example addressed the issue of route schedule optimization for the low-cost airline that travels between the five major US airports of Atlanta, Los Angeles, Chicago, Dallas, and New York. The case study indicates that the most effective method for making informed decisions in companies is to combine quantitative and qualitative decision-making techniques.

The optimal flight schedule has given the aforementioned criteria, and the corresponding value of profit, may be easily discovered using a Linear Programming solver tool. In our case study, this answer is represented as the best flight schedule. It is feasible to do an analysis of the data to confirm the sensitivity of the optimal solution to any modifications based on the outcomes acquired from the resolution of the Linear Programming issue. The name of this procedure is Sensitivity Analysis. Through this analysis, crucial information can be acquired that will help the decision-maker to understand the problem's degrees of freedom and provide them with the answers to hypothetical questions. Therefore, it turns out that the Sensitivity Analysis is a crucial instrument for the post-optimal evaluation of potential new situations, as seen in the case study [22].

It is vital to rely on analysis methodologies that help the decision-making process in high complexity and unpredictable situations, such as the transportation sector, especially when faced with options that must satisfy numerous objectives and the interests of many stakeholders. It is crucial to select the most effective approach to support decision-making based on the situation, the requirements, and the desired outcomes [23].

Furthermore, using mathematics and psychology, the Analytic Hierarchy Process has been described as a multi-criteria decision analysis process that may recommend the optimal solution from a limited number of options, weighed on the preferences of the decision maker. The technique assigns priority values to specific options, which are decided by a multilevel hierarchical structure that evaluates each option's significance using both quantitative and qualitative criteria. Analytic Hierarchy Process supports decision-makers in locating the option that most closely aligns with their beliefs and perception of the situation.

Contrasting the Analytic Hierarchy Process, Linear Programming offers an objective optimal solution based only on the final goal and the constraints on which the system is subject are unaffected by the decision maker's subjectivity. A preference criterion may sometimes be required to be inserted in order to choose among the many possibilities, making the Analytic Hierarchy Process preferred to Linear Programming.

Finally, the research described how the two models can be combined to produce an integrated Analytic Hierarchy Process-Linear Programming model (AHP-LP), which enables the inclusion of some flexibility in the search for the optimal solution, represented by the subjective weights attributed to the Analytic Hierarchy Process, in order to obtain more precise results and better adapt the quantitative decision support model to the problem. Analytic Hierarchy Process is used to create a priority scale based on the subjective assessments of decision-makers and is capable of evaluating both quantitative and qualitative elements. In order to discover the best solution that maximizes the intended benefits, the priority ratings acquired from the Analytic Hierarchy Process model are then employed as coefficients of the decision variables in a Linear Programming model.

8. Conclusion and future findings

Reviewing the definition of strategy, we can certainly recall that it has been seen by Chandler as the determination of long-term organizational goals, objectives, lines of conduct, and criteria for resource allocation [24].

In addition, the studies of Hofer and Schendel have highlighted how strategy means the identification of the means or rather the "system of current and planned use of resources and interaction with the environment" that the company plans to use to try to achieve its objectives [25].

It is then worth highlighting the centrality that Grant gives to the link between strategy and value, that is, arguing that strategy is concerned with success, or rather with guiding the decision-making process of the management toward the achievement of the success of the company or rather the creation of value.

Finally, strategy, in the view of Boschetti, is shown as an integrated set of decisions aimed at ensuring the company's competitive advantage over time and in comparison with competitors [26].

Taken as a whole, these definitions describe the complexity of the function of an organizational strategy, leading to the observation of a fundamental element, namely that strategy is the model of the pursuit of entrepreneurial success that the company has in fact adopted or intends to adopt, in order to excel in the competitive confrontation [27].

Strategy concerns both the choice of ends and the choice of means to achieve them because under conditions of limited rationality, it is not always possible to completely separate means from ends (Simon), strategy is sometimes improvisation and, finally, objectives can sometimes be discovered after starting to operate [28].

For this reason, possessing more decision-making elements makes it possible to direct decision-making processes in a more analytical manner.

Thus, reviewing what has been discussed, the literature certainly provides us with an analysis of how unavoidable decision-making is for organizations and how the correct decision, although complex to make, can determine the success of the company's objectives.

Sometimes, only qualitative decision-making processes are not enough to efficiently support the management in the definition of the operative scenario and in identifying business opportunities and possible solutions [29].

Hence, this paper later introduces some analytical and multi-criteria models, such as Linear Programming and Analytic Hierarchy Process, which allow us to reach apparently more correct and complete decisions.

Through the case study presented in the discussion section, it has been demonstrated the importance and the utility of Linear Programming in both planning and operational decision-making processes. In fact, this tool provides a numerical solution to the optimization problem, directly providing the best possible solution under a set of conditions. Without this instrument, an enormous human effort would be required to identify a solution that would not be so precise anyway.

With the aim of completing the analysis, the Analytic Hierarchy Process method has also been described, without however presenting any practical applications.

As discussed in the section above, it is possible to obtain an integrated Analytic Hierarchy Process-Linear Programming model to merge their best features into a single tool.

Further research may be conducted on this application of the mathematical models focusing on the possibility of creating a tool capable of supporting the decision-making process by solving problems of a different nature than the one analyzed in this paper.

Author details

Federico de Andreis*, Enza Curcio, Federico Maria Sottoriva and Ubaldo Comite
Università Giustino Fortunato, Benevento, Italy

*Address all correspondence to: f.deandreis@unifortunato.eu

References

[1] Comite U, de Andreis F, Mattavelli F. The evolution of aviation. Leading cases of efficient business models. European Journal of Economics, Finance and Administrative Sciences. 2021;**108**: 81-94

[2] Peterson GD, Cumming GS, Carpente SR. Scenario planning: A tool for conservation in an uncertain world. Conservation Biology. 2003;**17**(2): 358-366. DOI: 10.1046/j.1523-1739.2003.01491.x

[3] Le LG. infrastrutture e il turismo. Elementi di geografia dei trasporti. Milano: Vita e Pensiero; 2005

[4] Doganis R. Airline Business in the 21st Century. London: Routledge; 2000. DOI: 10.4324/9780203991916

[5] Terziev V, Georgiev M. Efficient management as optimization of the organization. IJASOS-International E-journal of Advances in Social Sciences. 2018;**4**(10):6-15. DOI: 10.18769/ijasos.415490

[6] Tversky A, Kahneman D. Rational choice and the framing of decisions. The Journal of Business. 1986;**59**(4):251-278

[7] Perlow LA, Okhuysen GA, Repenning NP. The speed trap: Exploring the relationship between decision making and temporal context. The Academy of Management Journal. 2002;**45**(5):931-955. DOI: 10.2307/3069323

[8] Zarghami M, Szidarovszky F. Introduction to multicriteria decision analysis. In: Zarghami M, Szidarovszky F, editors. Multicriteria Analysis. Berlin, Heidelberg: Springer; 2011. pp. 1-12. DOI: 10.1007/978-3-642-17937-2_1

[9] Steele K, Karmel Y, Cross J, Wilcox C. Uses and misuses of multicriteria decision analysis (MCDA) in environmental decision making. Risk Analysis: An International Journal. 2009; **29**(1):26-33. DOI: 10.1111/j.1539-6924.2008.01130.x

[10] Sena B. L'approccio del case study nella ricerca socio economica. Milano: Franco Angeli; 2021. DOI: 10.3280/SR2016-111001

[11] Kozlowsk SWJ, Ilgen DR. Enhancing the effectiveness of work groups and teams. Psychological Science in the Public Interest. 2006;**7**(3):77-124. DOI: 10.1111/j.1529-1006.2006.00030.x

[12] Kalinic I, Sarasvathy SD, Forza C. 'Expect the unexpected': Implications of effectual logic on the internationalization process. International Business Review. 2014; **23**(3):635-647. DOI: 10.1016/j.ibusrev.2013.11.004

[13] de Andreis F. A theoretical approach to the effective decision-making process. Open Journal of Applied Sciences. 2020; **10**:287-304. DOI: 10.4236/ojapps.2020.106022

[14] Bernoulli D. Exposition of a new theory on the measurement of risk. Econometrica. 1954;**22**(1):23-36. DOI: 10.2307/1909829

[15] Lichtenstein S, Slovic P. Response-induced reversals of preference in gambling: An extended replication in Las Vegas. Journal of Experimental Psychology. 1973;**101**:16-22

[16] Tversky A, Slovic P, Kahneman D. The causes of preference reversal. The American Economic Review. 1990; **80**(1):204-217

[17] Johnson-Laird P, Byrne R. Précis of deduction. Behavioral and Brain Sciences. 1993;**16**(2):323-333. DOI: 10.1017/S0140525X00030260

[18] Fagoyinbo IS, Akinbo RY, Ajibode IA, Olaniran Y. Maximization of profit in manufacturing industries using linear programming techniques: Geepee Nigeria Limited. Mediterranean Journal of Social Sciences. 2011;**2**(6):97-105

[19] Jansen B, de Jong JJ, Roos C, Terlaky T. Sensitivity analysis in linear programming: Just be careful! European Journal of Operational Research. 1997; **101**(1):15-28. DOI: 10.1016/S0377-2217(96)00172-5

[20] Saaty TL. A scaling method for priorities in hierarchical structures. Journal of Mathematical Psychology. 1997;**15**(3):234-281. DOI: 10.1016/0022-2496(77)90033-5

[21] Ghodsypour SH, O'Brien C. A decision support system for supplier selection using an integrated analytic hierarchy process and LP. International Journal of Production Economics. 1998; **56**(1):199-212

[22] Saltelli A, Tarantola S, Campolongo F, Ratto M. Sensitivity Analysis in Practice: A Guide to Assessing Scientific Models. Chichester: Wiley; 2004

[23] Bonjean Stanton MC, Roelich K. Decision making under deep uncertainties: A review of the applicability of methods in practice. Technological Forecasting and Social Change. 2021;**2021**:171. DOI: 10.1016/j.techfore.2021.120939

[24] Fligstein N. Chandler and the sociology of organizations. The Business History Review. 2008;**82**(2): 241-250

[25] Chrisman JJ, Hofer CW, Boulton WR. Toward a system for classifying business strategies. The Academy of Management Review. 1988; **13**(3):413-428. DOI: 10.2307/258089

[26] Boschetti C. Strategie e processi di pianificazione nelle imprese diversificate. Torino: Giappichelli; 1990

[27] Dean JW, Sharfman MP. Does decision process matter? A study of strategic decision-making effectiveness. The Academy of Management Journal. 1996;**39**(2):368-396. DOI: 10.2307/256784

[28] Estrada F. Economics and rationality of organizations: An approach to the work of Herbert A. Simon. SSRN Electronic Journal. 2010;**2010**. DOI: 10.2139/ssrn.1583199

[29] Hammarberg K, Kirkman M, de Lacey S. Qualitative research methods: When to use them and how to judge them. Human Reproduction. 2016;**31**(3): 498-501. DOI: 10.1093/humrep/dev334

Chapter 8

Predicting the Susceptibility to *Pityokteines curvidens* Using GIS with AHP and MaxEnt Models in Fir Forests

Fatih Sivrikaya, Gonca Ece Özcan and Korhan Enez

Abstract

One of the most damaging insects to coniferous trees, especially pine and fir, is *Pityokteines curvidens*. It may cause significant tree death and alter the composition and structure of the forest. Early detection of forest stands vulnerable to bark beetle infestations and can help reduce the harmful effects of possible infestations. The main objective of the study was to compare the susceptibility maps of *P. curvidens*, utilizing the analytical hierarchy process and maximum entropy approaches. The study was conducted in the fir forests of the Küre Forest Enterprise in Türkiye. *P. curvidens* susceptibility maps were created using the stand structure, forest stand form, site index, slope, aspect, elevation, NDVI, and solar radiation as the eight key criteria. The accuracy of the maps based on the analytical hierarchy process and maximum entropy approaches was assessed using the receiver operating characteristic curves. The most crucial variables in both models' predictions of *P. curvidens* susceptibility were found to be stand structure, NDVI, and elevation. Comparing the maps produced by the two models, the MaxEnt model gave more precise predictions than AHP. This study may make it easier for decision-makers to create susceptibility maps for fir engraver bark beetles.

Keywords: fir engraver beetle, Kazdağı fir, analytic hierarchy process, maximum entropy, susceptibility

1. Introduction

Bark beetles (Coleoptera: Curculionidae, Scolytinae), one of the most significant forest pests [1], are natural disturbances that have severe epidemics in the last century [2]. While only 2% of these species can cause epidemics in forests [3], endemic populations in small tree groups, and epidemic populations in large forest stand areas cause tree deaths [4, 5]. Host tree resistance prevents increasing of bark beetle populations [6]. However, attacks are easier on trees weakened by biotic and abiotic factors [7].

Epidemics significantly affects forest ecosystem values and causes deterioration of the functioning and characteristics of natural ecosystems [8, 9].

Abies (Miller) genus, which has two species and four subspecies in Türkiye [10], covers a total area of 511.703 ha [11]. Kazdağı fir (*Abies nordmanniana* ssp. *equi-trojani* (Asc. & Sint. ex Boiss.) Coode & Cullen), one of the endemic fir taxa for Türkiye, is economically and ecologically important and is very sensitive to climate variables [12]. Due to its declining population size and limited geographic range, the species has been given endangered status by the International Union for Conservation of Nature [13].

One of the most important of the bark beetles that damage fir is the fir engraver beetle (*Pityokteines curvidens* (Germar, 1824) Coleoptera: Curculionidae: Scolytinae) [14]. Although the main host of the species is Abies alba, it can also cause significant infestations in other Abies species (*A. equi-trojani, A. sibirica*, *A. cephalonica*, *A. balsamea*, *A. bornmülleriana*, *A. borisii-regis*, *A. cilicica*, *A. nordmanniana*, *A. firme*, *A. sachalinensis*, and *A. fraseri*). In addition, *Pinus silvestris*, *P. brutia*, *P. strobus*, *Pseudotsuga menziesii*, *Picea orientalis*, *P.abies*, *Larix decidua*, *L. kaempferi*, *Cedrus libani*, are other hosts [15, 16]. Abies ssp., Cedrus ssp., *Pinus nigra*, *P. brutia* and *Picea pungens* are hosts in Türkiye [17–20].

P. curvidens, one of the important pests of fir forests in the Western Black Sea region of Türkiye [21, 22], prefers forests and monoculture stands where suitable habitat conditions are not available [23]. Although it is generally a secondary pest, it also becomes a primary pest depending on the increase in its population [24, 25]. It has been recorded that bark beetles caused a total damage of 2,353,057 m^3 in 1987–2004 in this region [22].

Although it is not possible to prevent natural disasters, risks can be reduced by taking the necessary precautions [26]. Conducting risk assessment and modeling of the damages caused by bark beetles will help decision-makers in managing forest ecosystem [27]. In addition, there is a need for accurate and reliable estimation of pest distributions in forest ecosystems in today's forestry [28]. Many different modeling techniques are used to estimate the distributions of bark beetle. Models that predict the geographical distribution of species are important for various applications in the field of ecology and conservation [29, 30]. Maximum Entropy (MaxEnt), which is one of the machine learning methods, is widely preferred in modeling species distributions and habitat suitability [31]. Spatial distribution of *Dendroctonus mexicanus* [28], potential distribution of *D. ponderosae*, *D. brevicomis* [32], *Ips calligraphus*, *I. grandicollis* [33], *I. mannsfeldi* [34] and *I. aminutus* [35], according to current and future climate scenarios and tree death events caused by bark beetles were estimated by MaxEnt [36].

The analytic hierarchy process (AHP), one of the multicriteria decision analysis methods (MCDA), is a widely used method in solving complex decision-making problems in many different disciplines [37]. While AHP is used effectively in forestry [38–43], studies on bark beetles are limited [44–46].

In this study, AHP and MaxEnt approaches were used to predict the susceptibility of *P. curvidens* in Kazdağı fir forests. The accuracy of the models was calculated by the receiver operating characteristic (ROC) analysis. In the theory part of this book chapter, the bark beetle's infestation, their effects on the ecosystem, and the importance of bark beetle susceptibility maps, according to AHP and MaxEnt models were explained. In the material and methods section, the conceptual framework, dataset, model parameters, and models used were described. In the result and discussion section, the model results obtained according to two different models were evaluated and the results were compared with the literature. In the last section, current bottlenecks were mentioned and suggestions for models were presented.

2. Material and methods

2.1 Study area

Küre Forest Enterprise (FE), a case study area, is situated on the İsfendiyar mountain in the Kastamonu Regional Directorate of Forestry in the Western Black Sea region at 41° 42′ N and 33° 42′ E (**Figure 1**). 52% of the Küre FE (73694.9 ha) is covered with forest areas. The dominant tree species in the study area are Abies ssp., Fagus ssp., Quercus ssp., Pinus ssp., Alnus ssp., Carpinus ssp., and Ulmus ssp. Pure, and mixed fir forest stands in the study area is 29911.6 ha and 21% of the fir forests (6217.1 ha) were under attack by *Pityokteines curvidens*. In the study area, fir stands were infested intensively by *Pityokteines curvidens* between 2008 and 2017. The average elevation of the study area above sea level is 960 meters. The study area is in the Black Sea climate zone. The relative humidity level is high in all seasons. The weather is cool in summer and not very cold in winter, but rainy.

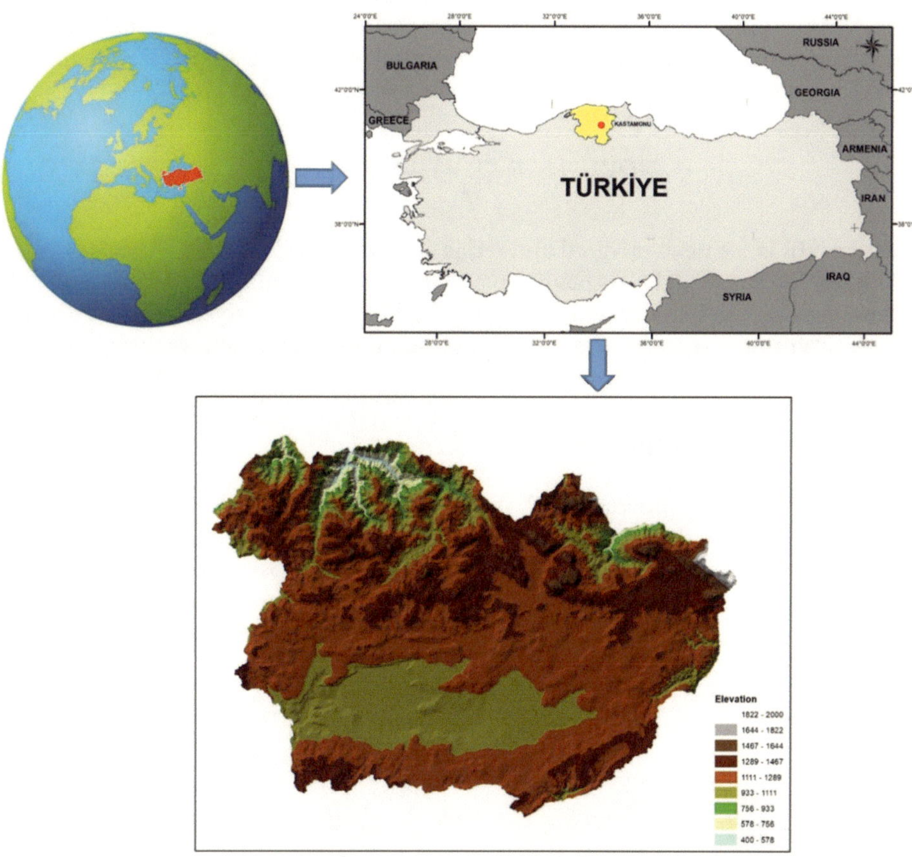

Figure 1.
The study area.

2.2 Parameters selection and dataset

A review of the literature helped determine the criteria that should be used to assess a forest stand's vulnerability to *Pityokteines curvidens* infestation. Eight parameters were thought to be essential indicators for forecasting *P. curvidens* epidemics: stand structure, forest stand form, site index (m), slope (%), aspect (degree), elevation (m), NDVI, and solar radiation (WH/m^2) [27, 45].

A digital forest stand map was acquired from the Küre FE [47] and was processed using ArcGIS 10.6 to create the database. The forest stand form, stand structure, and site index layers were created using forest stand maps. To evaluate the damage caused by *P. curvidens*, a field study was conducted in Küre FE between 2008 and 2017, the most intense period of damage. The beginning of an attack by *P. curvidens* on a tree can be noticed in the bark by scum emission coming from the reproductive galleries' entryways. The entrance holes have a 1.5 mm diameter. It usually takes about 2 months for the normal canopy color to change to pale yellow and then reddish. The trees then turn yellow and brown, and eventually die old or weakened trees [48, 49]. Tress that displayed these symptoms were regarded as infected and stands with infested trees were also considered as infested stands. The digital forest stand map was used to create the infested stand map once the *P. curvidens*-infested forest stands were added to the database.

The digital elevation model (DEM) and Landsat images (Landsat 7 ETM+ and Landsat 8 OLI between 2008 and 2017) were downloaded from the U.S. Geological Survey website [50]. DEM data were used to create the elevation, aspect, slope, and solar radiation maps, and Landsat images were used to compute Normalized Difference Vegetation Index (NDVI). The following formula was used to calculate the NDVI:

$$NDVI = (NIR - Red) / (NIR + Red) \tag{1}$$

where NIR is the near-infrared and red is the red band.

The resolutions of all layers have to match to provide results that are more reliable. Therefore, using the bilinear resampling method in Spatial Analyst, the NDVI map at 30 m resolution was transformed to 25 m resolution. **Table 1** shows the spatial database for the case study area.

2.3 Analytical hierarchy process (AHP)

Analytical hierarchy process (AHP) is one of the multicriteria decision analyses [51, 52]. AHP analyzes descriptive, qualitative, and quantitative parameters with pairwise comparison and weighting. In AHP, parameters are compared with each other in pairs, a pairwise comparison matrix is created, and then the weights of these parameters are determined [53]. The AHP allows the interpretation of experience and knowledge by

Data description	Source	Data type	Resolution
DEM	USGS	Raster	25 m
Landsat 7 ETM+ and Landsat 8 OLI	USGS	Raster	30 m
Forest Stand Map	Küre FE	Vector	25 m

Table 1.
The data and its sources used in the study.

Importance scale	Definitions of importance	Description
1	Equal	The objective is equally benefited by two actions.
3	Moderate	Judgment and experience are slightly preferred to one another.
5	Strong	Judgment and experience are strongly preferred to one another.
7	Very strong	Judgment and experience are very strongly preferred to one another.
9	Extreme	Knowledge and judgment are the strongest conceivable arguments.
2, 4, 6, 8	Intermediate	When a compromise is necessary.

Table 2.
The description of pairwise comparison in AHP.

modeling not only the basic parameters but also the relationship of the sub-parameters with each other in a certain hierarchical structure. These features make the method more flexible, and its systematic, hierarchical, and adaptive features make the AHP method very popular nowadays, where environmental problems are constantly increasing. For this reason, it is used effectively in various disciplines, such as landslide mapping [54], flood hazard mapping [53], forest fire risk mapping [55, 56], and bark beetle mapping [45, 57]. The knowledge of certain specialists is used to evaluate the comparative magnitudes of variables through pairwise comparisons at the first stage in AHP. Coefficients ranging from 1/9 to 9 are given to compare the relative importance of parameters for each pair in pairwise comparison (**Table 2**) [54, 58]. Consistency Ratio (CR) is an important validation parameter for the judgment matrix in AHP. The Random Index (RI) should be known before proceeding to the calculation of the CR. RI values were gathered from Saaty [52, 58] in this study. CR was calculated according to Eqs. (2) and (3) to determine the weight of the parameters and to demonstrate their consistency. If the CR is <0.10, it can be said that the generated judgment matrix is consistent. It is necessary to reevaluate the pairwise comparison matrix when the CR is >0.1.

$$CI = \lambda_{\max} - n / n - 1 \tag{2}$$

$$CR = CI / RI \tag{3}$$

where λ_{max} is the biggest value, n is the number of criteria, and CI is the consistency index.

2.4 Maximum entropy modeling (MaxEnt)

Maximum entropy (MaxEnt) is one of the machine learning methods and estimates the probability distribution of species habitats using presence-only data [59]. MaxEnt uses environmental variables as the independent variable and the distribution points of the species as the independent variable. MaxEnt 3.4.4 program was used to determine the potential distribution of fir engraver bark beetle [30, 59, 60]. To create the MaxEnt model, first of all, data sets were created. In this context, stand structure, forest stand form, site index, slope, aspect, elevation, NDVI, and solar radiation were used as variables. These variables have been converted to ASCII format. Stand structure, forest stand form, and site index were considered categorical data, while the others were considered continuous data. The jackknife test was used to determine the contribution rates of the variables in the model to be developed, and the AUC was used to evaluate the model performance. Implementation of the cross-validation method included 10 repetitions, a maximum of 5000 iterations, 1 regularization multiplier, and retention of other parameters at their default values. The logistic output grid format was chosen.

2.5 Fir engraver beetle susceptibility mapping

The risk of a bark beetle infestation is estimated by the fir engraver beetle susceptibility index (FEBSI). Infestation risk is increased by a high FEBSI and *vice versa*. The significance of each parameter in the AHP model was added up to estimate the FEBSI. All parameter layers were overlaid and the FEBSI values were computed in the AHP model using Raster Calculator tool in ArcGIS 10.6. FEBSI was calculated automatically, according to the selected parameters in MaxEnt model. Based on AHP and MaxEnt models, the susceptibility maps for the fir engraver bark beetle were created. FEBSI were categorized into four groups as low, moderate, high, and extreme based on the natural breaks' categorization technique in ArcGIS. Two fir engraver beetle susceptibility maps were created according to AHP and MaxEnt models.

2.6 Validation of fir engraver beetle susceptibility mappings

One of the most important stages of predictive models is assessing their performance. Therefore, the accuracy of the AHP and MaxEnt-based beetle susceptibility maps was assessed with the Receiver Operating Characteristic (ROC) method in this study. One of the most popular methods for verifying model-based forecasting is the ROC curve analysis, which ranks the precision of probabilistic and deterministic models, as well as compares them [42]. It is a graphic technique for investigating the tradeoff between specificity and sensitivity with the x-axis and the y-axis, respectively. The area under the curve (AUC) is a commonly used indicator of prediction accuracy in ROC curve analysis. The best outcome is to have the highest AUC score, which varies between 0 and 1. An AUC score of 1 indicates a perfect prediction of the model. Five categories are used to categorize the AUC score: poor (0.5–0.6), medium (0.6–0.7), good (0.7–0.8), very good (0.8–0.9), and perfect (0.9–1.0) [33, 42, 45, 61–64]. For the development of the model, 75% of the infested stands were utilized as training data, while the remaining 25% were used as validation data.

3. Results and discussion

Fir engraver beetle susceptibility maps were produced in a GIS environment using AHP and MaxEnt models. Many different parameters that affect the sensitivity of the beetle have been taken into account to prepare susceptibility maps. The CR was calculated as 0.053 and all pairwise comparison matrices were found to be consistent and satisfactory in the AHP method (**Table 3**). Stand structure, NDVI, solar radiation, and elevation were found to be more important than other parameters for fir engraver beetle susceptibility with weights of 2.431, 2.354, 1.443, and 1.315, respectively.

The percent contribution and permutation importance results of the estimation parameters based on the Jackknife test were given in **Table 4** and **Figure 2**. Parameters

Factors	Weights	CR	Classes	Weights	CR
Stand structure	2.431	0.053	Pure	0.667	0.085
			Mixed	0.333	
Forest stand form	0.555		Even-aged	0.250	0.073
			Uneven-aged	0.750	
Site index (m)	0.689		>37.6	0.091	0.023
			28.3–37.6	0.160	
			23.3–28.3	0.203	
			17.0–23.3	0.245	
			<17.0	0.301	
Aspect	0.778		North	0.412	0.098
			West	0.281	
			East	0.191	
			South	0.116	
Slope (%)	0.435		0–20	0.312	0.046
			20–40	0.490	
			>40	0.198	
Elevation (m)	1.315		<1000	0.122	0.089
			1000–1200	0.333	
			1200–1400	0.390	
			>1400	0.155	
NDVI	2.354		<0.25	0.089	0.079
			0.25–0.35	0.144	
			0.35–0.45	0.395	
			>0.45	0.372	
Solar radiation (WH/m^2)	1.443		<750,000	0.140	0.054
			750,000–800,000	0.177	
			800,000–850,000	0.264	
			>850,000	0.419	

Table 3.
The weights and consistency ratios for all variables in AHP.

were ranked based on percentage contribution as stand structure (74.4%), site index (10.9%), elevation (5.4%), and NDVI (2.6%). According to these results, stand structure, NDVI, and elevation were determined as the most important parameters in fir engraver beetle susceptibility in both models. This situation reveals that the stand structure, NDVI, and elevation should be taken into account while preparing the beetle risk map.

According to the ROC analysis for AHP model, the accuracy of the fir engraver beetle susceptibility mapping was calculated as 0.710, indicating a good prediction of the model. In the MaxEnt model, satisfactory results were also obtained for the fir engraver beetle susceptibility mapping with an AUC value of 0.739, which is higher

Variable	Percent contribution	Permutation importance
Stand structure	74.4	65.7
Site index	10.9	8.2
Elevation	5.4	7.6
NDVI	2.6	4.3
Solar radiation	2.5	6.9
Slope	2.3	4.8
Aspect	1.6	2.3
Forest stand form	0.3	0

Table 4.
Selected parameters and their percentage contribution.

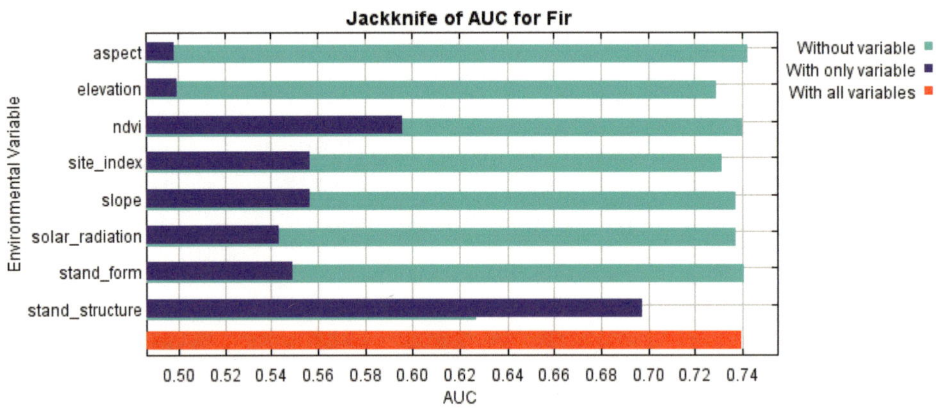

Figure 2.
The Jackknife test of selected parameters for fir engraver beetle.

than 0.5 of a random model (**Figure 3**). The maps produced according to both models gave reliable results. When both models were compared, it was revealed that the MaxEnt model produced more precise forecasts than AHP method. It can say that both AHP and MaxEnt give consistent results when preparing a beetle risk map.

Fir engraver beetle susceptibility maps were created according to AHP and MaxEnt models (**Figure 4**). According to the AHP method, 5.1%, 14.2%, 27.4%, and 53.3% of the study area are low, moderate, high, and extreme susceptibility classes (**Table 5**). To verify the results, the maps of infested forest stands and fir engraver beetle susceptibility overlapped. It showed that 80.7% of the infested stands were high and extreme susceptibility classes. According to the MaxEnt model, 40.0% of the study area was in moderate susceptibility classes, while 28.5% is extreme susceptibility classes. 60.5% and 12.8% of infested stands were in extreme and high susceptibility classes, respectively. According to both models, fir forest stands are susceptible to fir engraver beetle.

Our results are compatible with other studies. González-Hernández et al., [28] estimated the geographic distribution of Mexican pine bark beetle (*Dendroctonus mexicanus*) in conifer forests in Mexico using the MaxEnt model. They used bioclimatic (temperature, precipitation, etc.), and topographic variables (altitude, slope, and aspect) for bark beetle mapping. The potential distribution of the bark beetle was

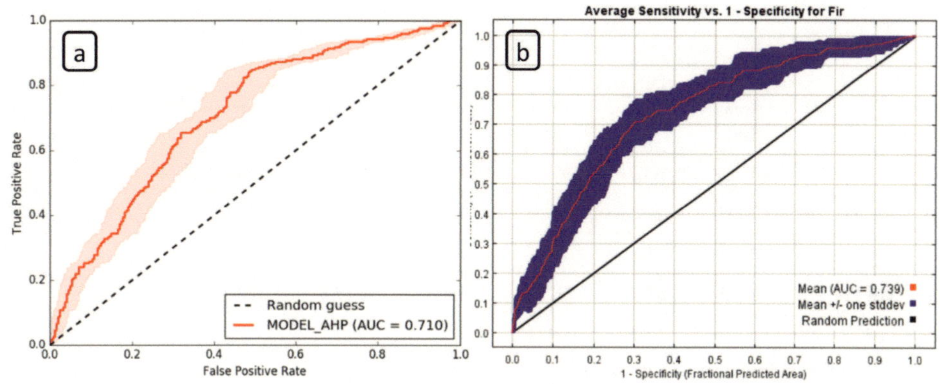

Figure 3.
The ROC curve of the fir engraver beetle susceptibility maps created by (a) AHP and (b) MaxEnt models.

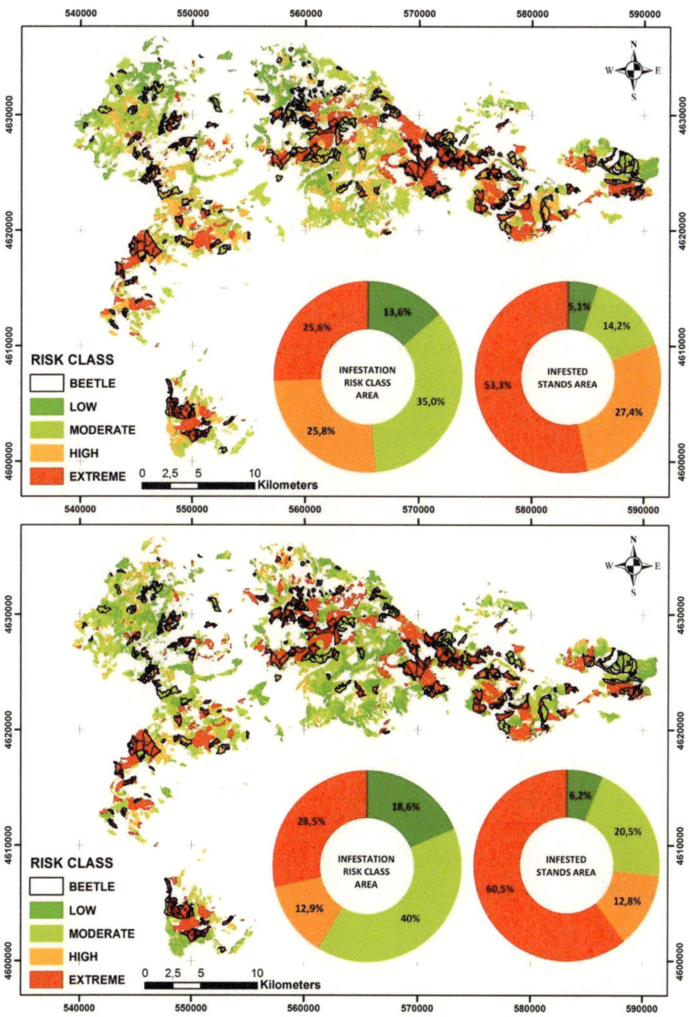

Figure 4.
Pityokteines curvidens susceptibility maps produced by (a) AHP and (b) MaxEnt models.

ISC	AHP				MaxEnt			
	SA		IS		SA		IS	
	Area (ha)	**%**	**Area (ha)**	**%**	**Area (ha)**	**%**	**Area (ha)**	**%**
Low	4054.5	13.6	315.9	5.1	5576.9	18.6	384.3	6.2
Moderate	10473.5	35.0	882.5	14.2	11952.8	40.0	1272.7	20.5
High	7729.7	25.8	1706.6	27.4	3852.1	12.9	796.6	12.8
Extreme	7653.9	25.6	3312.1	53.3	8529.8	28.5	3763.5	60.5
Total	29911.6	100	6217.1	100	29911.6	100	6217.1	100

ISC: infestation susceptibility class, SA: study area, and IS: infested stands.

Table 5.
The class areas in susceptibility maps based on the AHP and MaxEnt models.

reliable, according to the AUC (0.93). The key component in identifying suitable bark beetle habitats was found to be stand structure. Sarıkaya et al. [34] tried to model the current and future habitats of *Ips mannsfeldi* in Türkiye using MaxEnt. RCP4.5 and RCP8.5 climate change scenarios, as well as bioclimatic factors were used to estimate bark beetle habitats. They stated that elevation was the most important variable and estimated habitat distribution showed high accuracy with an AUC (0.88). Sivrikaya et al., [45] produced *I. sexdentatus* susceptibility map based on AHP, Frequency Ratio (FR), and Logistic Regression (LR) models. They stated that crown closure was the most important variable for three models. The accuracy of *I. sexdentatus* susceptibility maps for the LR, FR, and AHP models `was evaluated using the ROC curve, and the FR model provided a more precise estimate than the others.

4. Conclusion

The invasion of bark beetles is influenced by a variety of biotic and abiotic variables. Early detection of forest stands vulnerable to bark beetle infestations can help reduce the harmful effects of possible infestations. Thus, it is ensured that more accurate decisions are taken for future of forests in the understanding of sustainable management. In this study, susceptibility maps of *Pityokteines curvidens* were created using analytical hierarchy process (AHP) and maximum entropy (MaxEnt) approaches and their accuracy was compared. According to these results, stand structure, NDVI, and elevation were determined as the most important parameters in fir engraver beetle susceptibility in both models. Comparing the maps created from both models, the MaxEnt model produced more accurate predictions than the AHP. However, it was seen that susceptibility maps based on AHP have also acceptable accuracy. After all, according to both models, fir forest stands are susceptible to the fir engraver beetle and the key component in identifying suitable bark beetle habitats was found to be stand structure.

Acknowledgements

We would like to say thank you to the Küre Forest Enterprise for providing the forest stand map and related fir engraver beetle information.

Author details

Fatih Sivrikaya*, Gonca Ece Özcan and Korhan Enez
Faculty of Forestry, Department of Forestry Engineering, Kastamonu University, Kastamonu, Turkey

*Address all correspondence to: fsivrikaya@kastamonu.edu.tr

References

[1] Raffa KF, Grégoire J-C, Lindgren BS. Natural history and ecology of bark beetles. In: Vega FE, Hofstetter RW, editors. Bark Beetles- Biology and Ecology of Native and Invasive Species. Academic Press: Elsevier; 2015. pp. 1-40

[2] Negrón JF, Huckaby L. Reconstructing historical outbreaks of mountain pine beetle in lodgepole pine forests in the Colorado Front Range. Forest Ecology and Management. 2020;**473**:118270. DOI: 10.1016/j.foreco.2020.118270

[3] Bentz BJ, Jönsson AM. Modeling bark beetle responses to climate change. In: Vega FE, Hofstetter RW, editors. Bark Beetles- Biology and Ecology of Native and Invasive Species. Academic Press: Elsevier; 2015. pp. 533-553

[4] Sibold JS, Veblen TT, Chipko K, Lawson L, Mathis E, Scott J. Influences of secondary disturbances on lodgepole pine stand development in Rocky Mountain National Park. Ecological Applications. 2007;**17**:1638-1655. DOI: 10.1890/06-0907.1

[5] Raffa KF, Aukema BH, Bentz BJ, Carroll AL, Hicke JA, Turner MG, et al. Cross-scale drivers of natural disturbances prone to anthropogenic amplification: The dynamics of bark beetle eruptions. Bioscience. 2008;**58**:501-517. DOI: 10.1641/B580607

[6] Boone CK, Aukema BH, Bohlmann J, Carroll AL, Raffa KF. Efficacy of tree defense physiology varies with bark beetle population density: A basis for positive feedback in eruptive species. Canadian Journal of Forest Research. 2011;**41**:1174-1188. DOI: 10.1139/x11-041

[7] Faccoli M, Battisti A, Masutti L. Phenology of *Tomicus destruens* (Wollaston) in northern Italian pine stands. In: Lieutier F, Ghaioule D, editors. Entomological Research in Mediterranean Forest Ecosystems. Morocco: INRA; 2005. pp. 185-193

[8] Wermelinger B. Ecology and management of the spruce bark beetle *Ips typographus*—A review of recent research. Forest Ecology and Management. 2004;**202**:67-82. DOI: 10.1016/j.foreco.2004.07.018

[9] Kapil R, Marvasti-Zadeh SM, Goodsman D, Ray N, Erbilgin N. Classification of Bark Beetle-Induced Forest Tree Mortality using Deep Learning. 2022. arXiv preprint arXiv:2207.07241

[10] Akkemik Ü, Oral D. Türkiye'nin doğal gymnospermleri (Açık Tohumlular). OGM Yayınları. 2011;**2011**:214

[11] GDF. Türkiye Orman Varlığı. [Internet]. 2020 . Available from: https://www.ogm.gov.tr/tr/ormanlarimiz-sitesi/TurkiyeOrmanVarligi/Yayinlar/2020%20T%C3%BCrkiye%20Orman%20Varl%C4%B1%C4%9F%C4%B1.pdf [Accessed: November 26, 2022]

[12] Kara F, Lhotka JM. Climate and silvicultural implications in modifying stand composition in mixed fir-pine stands. Journal of Sustainable Forestry. 2020;**39**:511-525. DOI: 10.1080/10549811.2019.1686030

[13] IUCN. International Union for Conservation of Nature. [Internet]. 2022. Available from: https://www.iucnredlist.org/species/31325/9626365. [Accessed: October 15, 2022]

[14] Sarikaya O, Avci M. Predators of scolytinae (Coleoptera: Curculionidae) species of the coniferous forests in

the Western Mediterranean Region, Turkey. Turkish Journal of Entomology. 2009;**33**:253-264

[15] Wood SL, Bright DE Jr. A catalog of Scolytidae and Platypodidae (Coleoptera), part I: Bibliography. Great Basin Naturalist Memoirs. 1987;**1987**:11

[16] Pennacchio F, Gatti E, Roversi PF. Attacchi di Pityokteines spp. (Coleoptera Scolytidae) su *Abies alba* Miller in abetine dell'Appennino settentrionale e dell'Aspromonte. In: Proceedings of the Atti XIX Congresso nazionale italiano di Entomologia. Catania; 2002. pp. 921-925

[17] Schimitschek E. Forest Insects of Turkey and their Environment. Basics of Turkish Forest Entomology. Prag: Volk and Reich Verlag; 1944. p. 371

[18] Tosun İ. Insect pests of the coniferous forests of the Mediterranean region and researches on parasites and predators of their important species. Forestry General Directorate Publications. 1975;**612**:201

[19] Sekendiz OA. *Abies nordmanniana* Sıev (Spach.ı'ın Doğu Karadeniz bölümü ormanlarındaki zararlı böcekleri ile koruma ve savaş). Trabzon; 1982. p. 201

[20] Yüksel B, Öztürk N. First record of the silver fir bark beetle *Pityokteines curvidens* (Germ.) on blue spruce (Picea pungens Engelm.) in Turkey. Journal of Bartin Faculty of Forestry. 2022;**24**:25-31. DOI: 10.24011/barofd.1010243

[21] Toper KA. The Biology of Pityokteines Curvidens (Germ.) (Coleoptera, Scolytidae) damaged on fir in Bartın and Karabük forests. Zonguldak: Zonguldak Karaelmas University; 1999

[22] Serin M, Erdem M, Yüksel B, Akbulut S. Determination of the life cycle of effective bark beetles at Bolu and Aladağ Fir (*Abies bornmülleriana* Mattf.) stands and their possible preventive measurements. The Western Black Sea Forestry Research Institute. 2005;**12**:84

[23] Chararas C. The biology of *P. curvidens* and its behaviour with oleoresin extractives. Annales de l'Institut National Agronomique, Alencon. 1958;**44**:83-131

[24] Schwerdtfeger F. Die Waldkrankheiten: Lehrbuch der Forstpathologie und des Forstschutzes. Waldkrankheiten. 4th ed. Berlin: Hamburg; 1981

[25] Urban J. Diagnostics of bark beetles of the genus Pityokteines Fuchs important in forestry. Journal of Forest Science. 2002;**48**:329-341. DOI: 10.17221/11925-JFS

[26] Javidan N, Kavian A, Pourghasemi HR, Conoscenti C, Jafarian Z, Rodrigo-Comino J. Evaluation of multi-hazard map produced using MaxEnt machine learning technique. Scientific Reports. 2021;**11**:1-20. DOI: 10.1038/s41598-021-85862-7

[27] Özcan GE, Sivrikaya F, Sakici OE, Enez K. Determination of some factors leading to the infestation of *Ips sexdentatus* in crimean pine stands. Forest Ecology and Management. 2022;**519**:120316. DOI: 10.1016/j.foreco.2022.120316

[28] González-Hernández A, Morales-Villafaña R, Romero-Sánchez ME, Islas-Trejo B, Pérez-Miranda R. Modelling potential distribution of a pine bark beetle in Mexican temperate forests using forecast data and spatial analysis tools. Journal of Forestry Research. 2020;**31**:649-659. DOI: 10.1007/s11676-018-0858-4

[29] Graham CH, Ferrier S, Huettman F, Moritz C, Peterson T. New developments in museum-based informatics and

applications in biodiversity analysis. Trends in Ecology & Evolution. 2004;**19**:497-503. DOI: 10.1016/j.tree.2004.07.006

[30] Phillips SJ, Dudík M. Modeling of species distributions with maxent: New extensions and a comprehensive evaluation. Ecography. 2008;**31**:161-175. DOI: 10.1111/j.0906-7590.2008.5203.x

[31] Cunze S, Tackenberg O. Decomposition of the maximum entropy niche function–a step beyond modelling species distribution. Environmental Modelling & Software. 2015;**72**:250-260. DOI: 10.1016/j.envsoft.2015.05.004

[32] Evangelista PH, Kumar S, Stohlgren TJ, Young NE. Assessing forest vulnerability and the potential distribution of pine beetles under current and future climate scenarios in the Interior West of the US. Forest Ecology and Management. 2011;**262**:307-316. DOI: 10.1016/j.foreco.2011.03.036

[33] Li Y, Johnson AJ, Gao L, Wu C, Hulcr J. Two new invasive Ips bark beetles (Coleoptera: Curculionidae) in mainland China and their potential distribution in Asia. Pest Management Science. 2021;**77**:4000-4008. DOI: 10.1002/ps.6423

[34] Sarikaya O, Karaceylan IB, Sen I. Maximum entropy modeling (maxent) of current and future distributions of Ips mannsfeldi (Wachtl, 1879) (Curculionidae: Scolytinae) in Turkey. Applied Ecology and Environmental Research. ;**2018**:2527-2535. DOI: 10.15666/aeer/1603_25272535

[35] Økland B, Flø D, Schroeder M, Zach P, Cocos D, Martikainen P, et al. Range expansion of the small spruce bark beetle Ips amitinus: A newcomer in northern Europe. Agricultural and Forest Entomology. 2019;**21**:286-298. DOI: 10.1111/afe.12331

[36] Jaime L, Batllori E, Margalef-Marrase J, Navarro MÁP, Lloret F. Scots pine (*Pinus sylvestris* L.) mortality is explained by the climatic suitability of both host tree and bark beetle populations. Forest Ecology and Management. 2019;**448**:119-129. DOI: 10.1016/j.foreco.2019.05.070

[37] Wong JK, Li H. Application of the analytic hierarchy process (AHP) in multi-criteria analysis of the selection of intelligent building systems. Building and Environment. 2008;**43**:108-125. DOI: 10.1016/j.buildenv.2006.11.019

[38] Nasırı M. Determining the priority of effective factors on forest fire from analytical hierarchy process. Journal of Applied Biological Sciences. 2013;**7**:52-55

[39] Akay AO, Demir M, Akgul M. Assessment of risk factors in forest road design and construction activities with fuzzy analytic hierarchy process approach in Turkey. Environmental Monitoring and Assessment. 2018;**190**(9):1-12. DOI: 10.1007/s10661-018-6948-0

[40] Tiwari A, Shoab M, Dixit A. GIS-based forest fire susceptibility modeling in Pauri Garhwal, India: A comparative assessment of frequency ratio, analytic hierarchy process and fuzzy modeling techniques. Natural Hazards. 2021;**105**:1189-1230. DOI: 10.1007/s11069-020-04351-8

[41] Kumari B. Pandey AC geo-informatics based multi-criteria decision analysis (MCDA) through analytic hierarchy process (AHP) for forest fire risk mapping in Palamau Tiger Reserve, Jharkhand state, India. Journal of Earth System Science. 2020;**129**:1-16. DOI: 10.1007/s12040-020-01461-6

[42] Sivrikaya F, Küçük Ö. Modeling forest fire risk based on GIS-based analytical hierarchy process and

statistical analysis in Mediterranean region. Ecological Informatics. 2022;**68**:101537. DOI: 10.1016/j.ecoinf.2021.101537

[43] Vijith H, Dodge-Wan D. Modelling terrain erosion susceptibility of logged and regenerated forested region in northern Borneo through the Analytical Hierarchy Process (AHP) and GIS techniques. Geoenvironmental Disasters. 2019;**6**:1-18. DOI: 10.1186/s40677-019-0124-x

[44] Falťan V, Petrovič F, Gábor M, Šagát V, Hruška M. Mountain landscape dynamics after large wind and bark beetle disasters and subsequent logging—Case studies from the Carpathians. Remote Sensing. 2021;**13**:3873. DOI: 10.3390/rs13193873

[45] Sivrikaya F, Özcan GE, Enez K, Sakici OE. Comparative study of the analytical hierarchy process, frequency ratio, and logistic regression models for predicting the susceptibility to *Ips sexdentatus* in Crimean pine forests. Ecological Informatics. 2022;**71**:101811. DOI: 10.1016/j.ecoinf.2022.101811

[46] Tahri M, Kašpar J, Madsen AL, Modlinger R, Zabihi K, Marušák R, et al. Comparative study of fuzzy-AHP and BBN for spatially-explicit prediction of bark beetle predisposition. Environmental Modelling. 2022;**147**:105233. DOI: 10.1016/j.envsoft.2021.105233

[47] FMP. Forest Management Plans of Küre Forest Enterprise. Ankara: General Directorate of Forestry; 2011

[48] Maksymov JK. Untersuchungen über den krummzähnigen Weißtannenborkenkäfer Ips curvidens Germ. während seiner Massenvermehrung 1947-49 in der Schweiz. Aus dem Entomologischen Institut der Eidg. Technischen Hochschule in Zürich. 1950;**1950**:499-581

[49] CABI. 2022. https://www.cabi.org/isc/datasheet/45720#toidentity

[50] USGS. Geological Survey. [Internet]. 2022 . Available from: https://earthexplorer.usgs.gov/ [Accessed: November 20, 2022]

[51] Saaty TL. The Analytic Hierarchy Process: Planning, Priority Setting and Resource Allocation. New York: McGraw-Hill; 1980

[52] Saaty TL. Theory and Applications of the Analytic Network Process. Decision Making with Benefits, Opportunities, Costs, and Risks. 3rd ed. RWS: Pittsburgh; 2005

[53] Kazakis N, Kougias I, Patsialis T. Assessment of flood hazard areas at a regional scale sing an index-based approach and analytical hierarchy process: Application in Rhodope–Evros region, Greece. Science of the Total Environment. 2015;**538**:555-563. DOI: 10.1016/j.scitotenv.2015.08.055

[54] Saha A, Mandal S, Saha S. Geo-spatial approach-based landslide susceptibility mapping using analytical hierarchical process, frequency ratio, logistic regression and their ensemble methods. SN Applied Sciences. 2020;**2**:1-21. DOI: 10.1007/s42452-020-03441-3

[55] Babitha BG, Danumah JH, Pradeep GS, Costache R, Patel N, Prasad MK, et al. A framework employing the AHP and FR methods to assess the landslide susceptibility of the Western Ghats region in Kollam district. Safety in Extreme Environments. 2022;**4**:171-191. DOI: 10.1007/s42797-022-00061-

[56] Bentekhici N, Bellal S, Zegrar A. Contribution of remote sensing and

GIS to mapping the fire risk of Mediterranean forest case of the forest massif of Tlemcen (north-West Algeria). Natural Hazards. 2020;**104**:811-831. DOI: 10.1007/s11069- 020-04191-6

[57] Rahaman H, Rehman S, Sajjad H. Indicator-based inherent forest vulnerability using multicriteria decision-making analysis in the Darjeeling District of West Bengal. In: Rani M, Chaudhary BS, Jamal S, Kumar P, editors. Towards Sustainable Natural Resources. Cham: Springer. pp. 51-67. DOI: 10.1007/978-3-031-06443-2_4

[58] Saaty TL. Decision Making for Leaders: The Analytic Hierarchy Process for Decisions in a Complex World. 3rd ed. Pittsburgh: RWS Publications; 2012

[59] Phillips SJ, Anderson RP, Schapire RE. Maximum entropy modeling of species geographic distributions. Ecological Modelling. 2006;**190**:231-259. DOI: 10.1016/j.ecolmodel.2005.03.026

[60] Phillips SJ, Anderson RP, Dudík M, Schapire RE, Blair ME. Opening the black box: An open-source release of maxent. Ecography. 2017;**40**:887-893. DOI: 10.1111/ecog.03049

[61] de Figueiredo M, Cordella CBY, Jouan-Rimbaud Bouveresse D, Archer X, Bégué JM, Rutledge DN. A variable selection method for multiclass classification problems using two-class ROC analysis. Chemometrics and Intelligent Laboratory Systems. 2018;**177**:35-46. DOI: 10.1016/j.chemolab.2018.04.005

[62] Duračiová R, Muňko M, Ivan B, Koreň M, Resnerová K, Holuša J, et al. A bark beetle infestation predictive model based on satellite data in the frame of decision support system TANABBO. iForest. 2020;**13**:215-223. DOI: 10.3832/ifor3271-013

[63] Gheshlaghi HA, Feizizadeh B, Blaschke B. GIS-based forest fire risk mapping using the analytical network process and fuzzy logic. Journal of Environmental Planning Management. 2020;**63**:481-499. DOI: 10.1080/09640568.2019.1594726

[64] Pourghasemi HR, Beheshtirad M, Pradhan B. A comparative assessment of prediction capabilities of modified analytical hierarchy process (M-AHP) and Mamdani fuzzy logic models using Netcad-GIS for forest fire susceptibility mapping. Geomatics, Natural Hazards Risk. 2016;**7**:861-885. DOI: 10.1080/19475705.2014.984247

Chapter 9

Definition of Analytical Hierarchy Process and Its Effect on Environmental Decisions in Engineering for Landfill Siting

Mahya Rostampoor

Abstract

The AHP technique was expanded by Saaty (1980) and called analytic hierarchy process. It is a usual way of multi-criteria decision-making (MCDM) since it was based on theoretical authority. It has a remarkable ability to solve complex problems throughout the decision-making process in any domain. This method, which mirrors human thinking and natural behavior, authorizes the decision-maker to present the interaction between varied criteria in complex and unstructured situations. The AHP method is employed to recognize the consistency of weightings for each criterion, which works by constructing a pairwise comparisons matrix. This chapter aims to examine this method in detail in a tangible example for the field of the environment (landfilling). To obtain this goal, an influential criterion that can affect the decision was considered by using geometric integration in Expert Choice with merging the AHP technique and GIS software. We used governmental regulations, expert's opinions and previous literature. Paveh is in (Kermanshah) province located in the western part of Iran with 736 km^2 area. The results show that almost 93% of the study areas are unsuitable for landfill sitting.

Keywords: AHP, MCDM, GIS, landfiill, expert choice

1. Introduction

A human being always faces various problems and issues in their life, which they have decide to solve or overcome. Identifying the most appropriate option that can design the path of life and determine its quality [1], swift economic and technological growth over the last fifty years has varied human lives and built modern society confront complex decision-making problems [2]. Multiple criteria decision-making (MCDM) is a significant part of contemporary decision science, aimed at supporting decision-makers faced with various decision criteria and multiple decision alternatives [3]. Considering the problems related to decision-making with multiple criteria, it can be said that due to the lack of standards, the speed and accuracy of decision-making decreases, and decision-making becomes difficult [4]. Such conditions make the decision-making process highly dependent on the decision-maker. To solve this problem or to minimize its side effects, multiple criteria decision-making is designed [5].

Forman and Selly believe that any multi-criteria decision-making system should have the following characteristics [6]:

- The possibility of formulating the problem and revising.
- Consider different options.
- Considering different indicators and criteria.
- The possibility of using qualitative and quantitative indicators in the decision-making process.
- Considering different people's opinions about options and indicators.
- The possibility of combining judgments.
- Having a strong theoretical basis.

An excellent and certified weight evaluation method is the analytical hierarchy process (AHP) [7]; this method is one of the most comprehensive processes designed for multi-factor decision-making because with this method, it is possible to formulate the problem hierarchically [8]. In this chapter, we will thoroughly present the AHP method and explain the analysis of this method in an example in the direction of landfilling and how to make a decision in this regard; we use the article of M. Rostampour et al. [9]. In this chapter, we will first explain the AHP process and how to use it, in the next step, how to design the matrix and weighting in this method and finally how to weight with Expert Choice software and integrate it with GIS and the result.

2. AHP process

The hierarchical process is one of the most comprehensive processes designed for multi-criteria decision-making because it is possible to deliberate different quantitative and qualitative criteria in the question. In this process, various options are absorbed in decision-making, and there is a possibility of sensitivity analysis on criteria and sub-criteria. Some other advantage of this multi-criteria decision-making method is determining the degree of compatibility and incompatibility of the decision. This method has a solid theoretical basis and was based on self-evident principles. Process AHP converts complex and challenging problems into a simple form and solves them [10, 11].

In this method, the decision-making problem was separated into varied levels:

- Goals: The central question of the research or the problem that is supposed to be solved is called the goal. The goal is the chief of the hierarchical diagram and has only one parameter [12].
- Criteria: The criteria that include the goal and its constituents are called criteria. The criteria are the touchstone of the goal or its measurement. As long as the criteria cover more target components and are more expressive of the target, the probability of getting a more accurate result will increase. The criteria are the second line of the hierarchical diagram after the goal. At this level, it is possible to divide and adjust the required number of criteria on a horizon level. Criteria can be divided into sub-criteria, sub-criteria can be divided into subsequent sub-criteria, and this situation can be increased up to n sub-criteria on vertical and horizontal levels depending on the necessity [13].

Alternative: The options are the purpose and destination of the target in the hierarchical diagram, and the target answer is feared and obtained among the options. Options are the last level of the hierarchical diagram and depend on the use of this diagram; in cases where this method is used for selection or prioritization, the options are generally determined by the researcher because they are the one who decides which options should be chosen or which options should be prioritized [14].

- Inclusion of quantitative and subjective criteria in the priority-setting process, as displayed in above **Figure 1**.

- This is the only MCDM technique that has an effective mechanism for checking the uniformity of weightage established by the decision-maker, and thus, it does not necessitate the decision-maker to be unnaturally consistent [15].

- AHP method prepares a scale for measuring qualitative criteria, provides a method for estimating priorities and finally improves the judgment by normalizing the weight of each criterion [16].

- In general, we take up to nine criteria with AHP, but we can include more criteria by separating criteria into sub-criteria [3].

3. The four principles used in method AHP

1. Inverse condition: If the preference of element A over element B is equal to n, then the preference of element B over element A will be equal to 1/n.

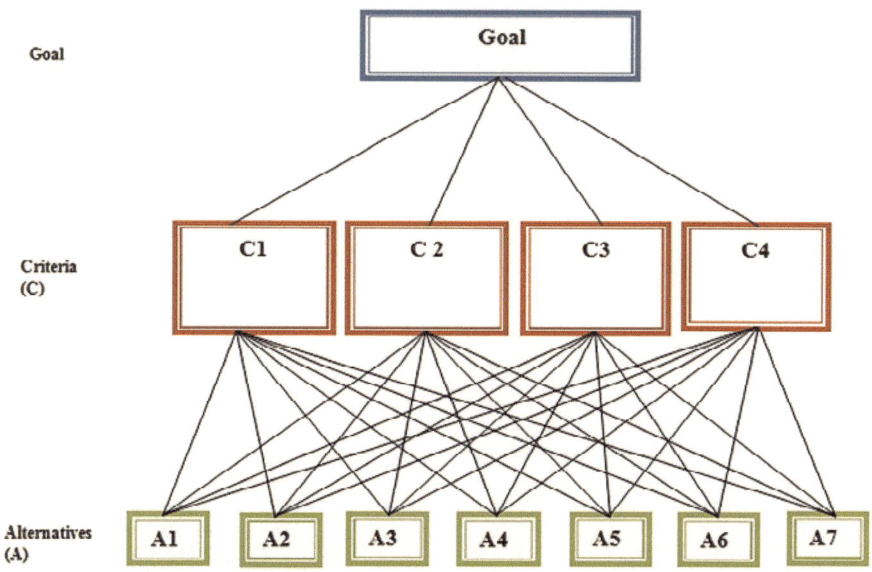

Figure 1.
AHP process.

2. Homogeneity: Element A and B must be homogeneous and comparable. In other words, the superiority of element A over element B cannot be infinite or zero.

3. Dependency: Every hierarchical element can depend on its higher-level element, and this dependence can continue up to the highest level linearly.

4. Expectations: Whenever there is a change in the hierarchical structure, the evaluation process must be done again [17].

To better understand the method, we use an environmental example and all its decision-making steps. In this example, which was done by the author in 2020, with the help of AHP & GIS, she identified the right place to build a landfill in Paveh county in Iran.

The first part of the course, for this example, is making a hierarchy for decision problems. The goal of the decision problem was landfill site selection for Paveh. The hierarchical construction was built based on the experts' opinions and earlier studies with convenient data in the study area. In this case, twelve criteria were elected and classified into the principal group. They are demonstrated in **Figure 2**. Group number one was environmental criteria that involve soil type, distance from the river, distance from fault, land use type, rainfall, distance from protected areas, wind direction and evaporation rate. Group number two was economic criteria including slope and

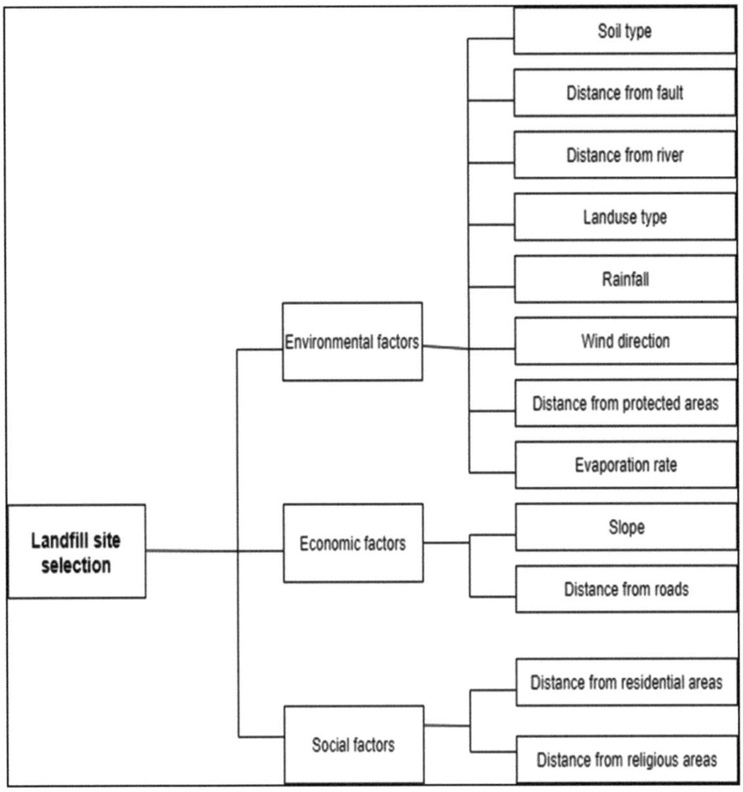

Figure 2.
Hierarchical structure.

distance from road sub-criteria, and group number three criteria were the social factors, which include distance from residential areas and distance from religious areas [9].

4. Formation of pairwise comparison matrices

At this stage, the elements of each level are compared to other related elements at a higher level in a pairwise manner, and pairwise comparison matrices were formed [18]. Allocation of numerical points related to the pairwise comparison of the importance of two options or two indicators is done based on **Table 1**.

A pairwise comparison matrix is show below:

$$\begin{bmatrix} a_{11} & a_{12} & \dots & a_{1n} \\ a_{21} & a_{22} & \dots & a_{2n} \\ \dots & \dots & \dots & \dots \\ a_{n1} & a_{n2} & \dots & a_{nn} \end{bmatrix} \tag{1}$$

where a_{ij} is the preference of element i over element j. In the pairwise comparison of the criteria with each other, the following relationship is established according to the inverse condition [19];

$$a_{ij} = \frac{1}{a_{ji.}} \tag{2}$$

The pairwise comparison matrix is an n × n matrix where n is the number of elements that have been compared [20].

For each pairwise comparison matrix n × n, elements on the diameter are equal to one, and there is no need to evaluate, but other parameters of the matrix must be determined based on pairwise comparisons.

The relative diameters are inverse to each other. Therefore, the number of pairwise comparisons for each n × n pairwise comparison matrix is equal to the [21]:

The Intensity of Importance	Description
1	Equal importance
2	Equal to average importance
3	Average importance
4	Average to strong importance
5	Strong importance
6	Strong to very strong importance
7	Very strong importance
8	Very strong or super strong importance
9	Super strong importance
2, 4, 6, 8	Intermediate values

Table 1
Saaty and Vargas preference scales [4].

$$N_c = \frac{n(n-1)}{2} \tag{3}$$

If the decision problem includes m options and n criteria, then n matrices m × m matrices and an n × n comparison matrix should be created. Therefore, the number of pairwise comparisons of the hierarchy (total problem) is equal to the

$$N_h = \frac{\mathrm{n}(\mathrm{n}-1)}{2} + \left(n \times \frac{m(m-1)}{2}\right) \tag{4}$$

Here, we discuss the pair of criteria comparisons that we designed in the landfill location questionnaire. These questions were asked to two experts in the field of environment and one person in Paveh Municipality and an educated native of that area, and finally, we measured their opinions together and obtained an average and entered the data into the Expert Choice software for review and weighting.

Question: In choosing the right place for the landfill center of Paveh city, is the criterion more important than your point of view? (**Table 2**)

Question: If only environmental and health criteria are considered in choosing the right place for the burial center, then which sub-criteria of the table below will be more important? (**Table 3**)

Choosing landfill place	Environmental factor	Economic factor	Social factor
Environmental factor	1		
Economic factor		1	
Social factor			1

Table 2.
Determining the weight of the main criteria

Environmental factor	Soil type	Distance from fault	Distance from river	Land use type	Rainfall	Wind direction	Distance from protected areas	Evaporation rate
Soil type	1							
Distance from fault		1						
Distance from river			1					
land use type				1				
Rainfall					1			
Wind direction						1		
Distance from protected areas							1	
Evaporation rate								1

Table 3.
Determining the weight of sub-criteria related to environmental and health criteria.

Economic factor	Slope	Distance from road
Slope	1	
Distance from road		1

Table 4.
Determining the weight of the sub-criteria related to the economic criterion.

Social factor	Distance from residential areas	Distance from religious areas
Distance from residential areas	1	
Distance from religious areas		1

Table 5.
Determining the weight of the sub-criteria related to the social criterion.

Question: If only economic criteria are considered in choosing the right place for the burial center, then which sub-criteria in the following table will be more important? (**Table 4**)

Question: If only social criteria are considered in choosing the right place for the burial center, then which sub-criteria of the following table will be more important? (**Table 5**)

5. Calculating the weight of elements in the hierarchical analysis method

In the hierarchical analysis method, the elements of each level are compared in pairs to each of the elements of the higher level, and their weights are computed [22]. These weights are called local priority (weight). Then, by combining the local priority, the overall priority of each option is determined. The weight of the criteria reflects their influence in determining the goal [19]. The weight of each option relative to the criteria is the portion of that option in the relevant criteria. Therefore, the overall priority of each option is obtained from the sum of the product of the weight of each criterion by the weight of the option of that criterion [23].

6. Calculation of local priority

After determining the pairwise comparison matrix, the local priority is calculated. There are different methods to calculate the local priority, which we mention here. You can research for more explanation: least squares method, logarithmic least square method, eigenvector method and approximation methods [24].

Approximation methods: This method will have fewer calculations than the previously mentioned methods, but they will also be less accurate. The most important methods of this method are 1—row sum, 2—column sum, 3—arithmetic mean 4—geometric mean [25].

In the first stage of this model, weighting was done based on the AHP method, for this aim, the criteria were weighted using the Delphi method and the opinion of experts from different fields, the criteria were weighted, and using the AHP method and software (Expert Choice11), Expert Choice software was used for the overall priority. Expert Choice software has many capabilities. One of its efficient cases is performing pairwise group comparisons and robust sensitivity analysis [26]. Pairwise group comparisons are used when more than one respondent is involved in decision-making, and sensitivity analysis is performed to check the weight of the options by changing the criteria. This software was developed by Thomas Saaty and Ernest Forman in 1983 by Expert Choice Inc. [6].

Video for how to work with Expert Choice: https://youtu.be/RzoDdms-6jc

The final weight of the layers was obtained with an inconsistency ratio of 0.05 (which is acceptable based on the Saaty theory, and we will explain this rate in full below). The distance layers from the fault and soil material had the furthest weight (**Figure** 3).

Inconsistency ratio: One of the advantages of the AHP is to control the consistency of the decision. In other words, A can calculate the compatibility of the decision and judge whether it is good, bad, acceptable or unacceptable. If A is twice as important as B and B is three times as important as C, if the importance of A compared to C is equal to 6, we say this judgment is consistent.

In general, if matrix A has inconsistency, we have the following theorems:

The sum of eigenvalues of matrix A is equal to n:

$$\sum_{i} \lambda_i = n \tag{5}$$

The eigenvalue of the matrix A is greater than or equal to the dimension of the matrix:

$$\lambda_{max} \geq n \tag{6}$$

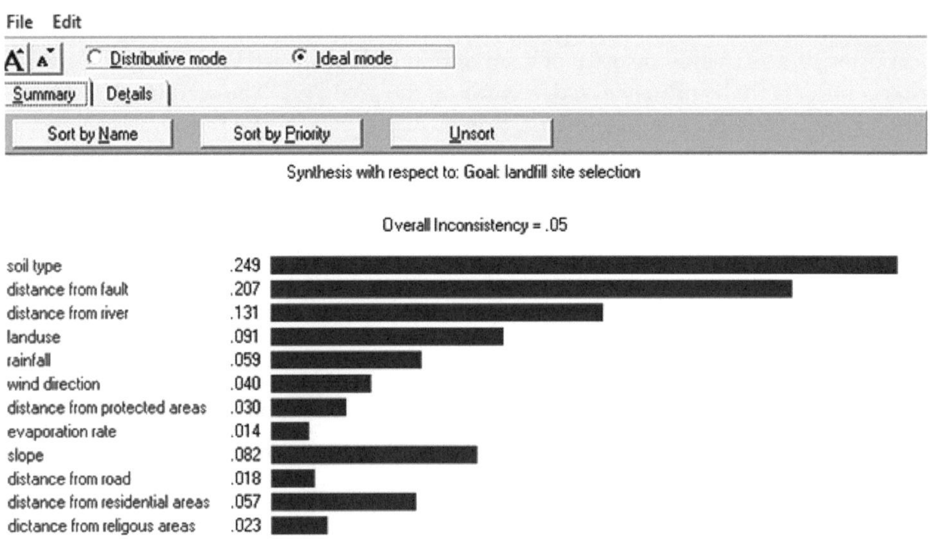

Figure 3.
Weighting on expert choice.

n	1	2	3	4	5	6	7	8	9	10	12
RI	0	0	0.58	0.9	1.12	1.24	1.32	1.45	1.45	1.51	1.53

Table 6
Values of (R.I.I) for the n-dimensional matrix.

When matrix A deviates slightly from the consistent state, λ_{max} will also deviate slightly from n. Therefore, the difference (λ_{max}-n) will be an excellent measure to measure the inconsistency of the matrix.

Undoubtedly, the scale ($\lambda_{max} - n$) depends on the value of n, and to solve this dependence, the scale can be defined as follows, which is called the inconsistency index [12].

$$I.I = \frac{\lambda_{max} - n}{n - 1} \tag{7}$$

They have calculated the value of the inconsistency index for the matrices whose numbers are entirely randomly chosen and called it the random inconsistency index. The values of (R.I.I) for the n-dimensional matrix can be obtained from the following table [27] (**Table 6**):

For each matrix, the result of dividing the inconsistency index by the random inconsistency index is called the inconsistency rate [28], which is a suitable criterion for judging inconsistency.

For each matrix, the result of dividing the inconsistency index by the random inconsistency index is called the inconsistency rate, which is a suitable criterion for judging inconsistency [29].

$$I.R = \frac{I.I}{R.I.I} \tag{8}$$

Calculating the inconsistency rate is also very important in the AHP method. In general, it can be said that the acceptable level of the inconsistency of a system depends on the decision-maker. Still, Saaty presents the number 0.1 as an acceptable limit and believes that if the amount of inconsistency is more than 0.1, it is better to revise the judgment [30].

Here, we will examine an example to understand the calculation of the inconsistency rate better. Then, we will go to the final solution of solving the general example and draw conclusions from it.

For the following pairwise comparison matrix, determine the weight vector from the arithmetic mean method and calculate its inconsistency rate.

	C_1	C_2	C_3
C_1	1	2	3
C_2	1/3	1	1/4
C_3	1/2	4	1

Solve: Arithmetic normalization is as follows: The elements of each column are added together, and each element is divided by this sum.

	C_1	C_2	C_3
C_1	0.545	0.375	0.6154
C_2	0.181	0.125	0.077
C_3	0.272	0.5	0.307

The weight vector comes from the linear average of the elements:

$$C_1 \cdot 0.512 \tag{9}$$

$$C_2 \cdot 0.128 \tag{10}$$

$$C_3 \cdot 0.360 \tag{11}$$

From the multiplication of the weight vector in the matrix, we have a pairwise comparison:

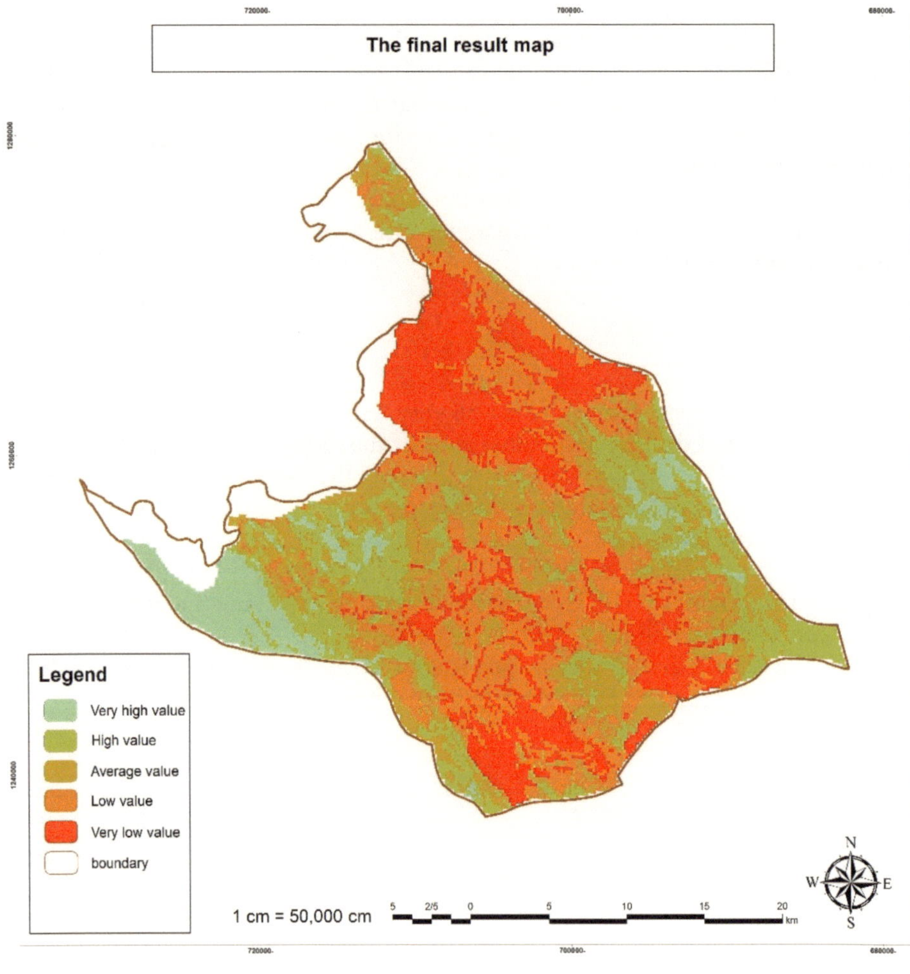

Figure 4.
Final map of landfill suitability.

$$\begin{vmatrix} 1 & 3 & 2 \\ 1/3 & 1 & 1/4 \\ 1/2 & 4 & 1 \end{vmatrix} \times \begin{vmatrix} 0.512 \\ 0.128 \\ 0.360 \end{vmatrix} = \begin{vmatrix} 1.616 \\ 0.388 \\ 1.128 \end{vmatrix} \quad (12)$$

In this case, the values of λ and λ_{max} are calculated as follows:

$$\lambda_1 = \frac{1.616}{0.512} = 3.156 \quad (13)$$

$$\lambda_2 = \frac{0.388}{0.128} = 3.036 \quad (14)$$

$$\lambda_3 = \frac{1.128}{0.360} = 3.133 \quad (15)$$

$$\lambda_{max} = \frac{3.156 + 3.0356 + 3.133}{3} = 3.108 \quad (16)$$

The inconsistency index

$$I.I = \frac{\lambda_{max} - n}{n - 1} = \frac{3.108 - 3}{3 - 1} = 0.054 \quad (17)$$

$$\text{R.I.I from } \textbf{Table 6} = 0.58 \quad (18)$$

$$I.R = \frac{I.I}{R.I.I} = \frac{0.054}{0.58} = 0.093 \quad (19)$$

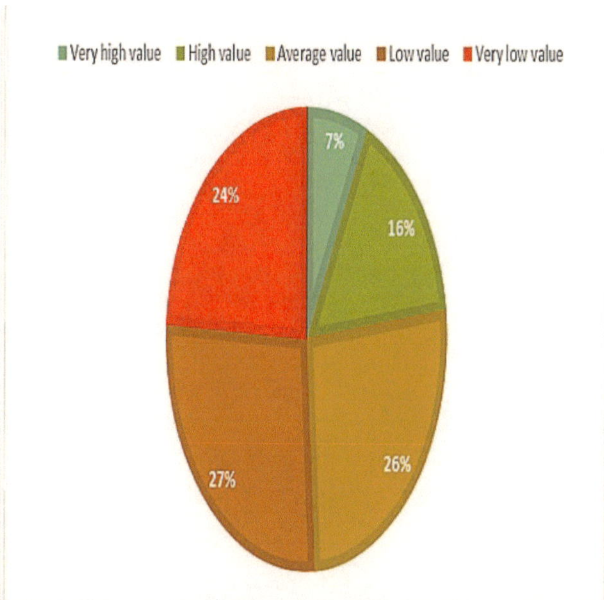

Figure 5.
Exhibition values with percentages.

7. Result and discussion

In the initial explanation of this chapter, we will further examine the AHP method based on an example previously worked by the author. The general results for the purpose of determining the location of the landfill for the desired area are shown in the form of a map below (**Figures 4–6**).

The weight of the layers must first be obtained; for this purpose, using Delphi method and the opinion of expert's in various fields, the criteria were weighted, and using AHP method and software (Expert Choice11), the final weight of the layers was calculated with an incompatibility coefficient of 0.05, in which the distance layers from the fault and soil material had the highest weight. In this chapter, an attempt was made to refer to the investigation of all ways of weighing and to understand it by giving a simple example.

In **Figure 6**, the final results of this investigation are shown with the help of GIS software and categorizing all the conditions from high to low value.

The result in **Figure 5** shows that 27% has poor landfill (orange), 26% has medium landfill (yellow), 24% of the study area has very poor landfill (red), 16% has good landfill (light green), and 7% has a very high capacity (dark green) for landfilling.

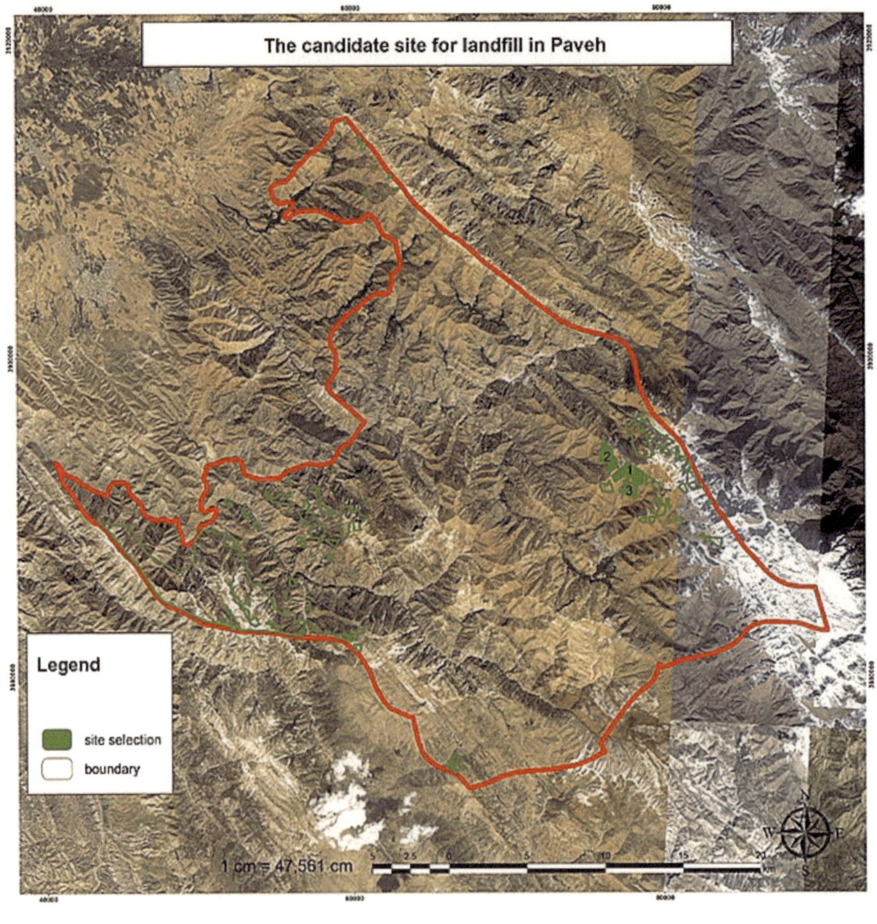

Figure 6.
Final landfill suitability map.

8. Conclusion

In this example, the full explanation of the AHP process, the selection method for weighting and the design of the questionnaire and reaching the right answer using the Expert Choice software and its integration with GIS were discussed. GIS is a powerful software that can provide rapid and perfect evaluation and has a great ability to manage massive volumes of data from different sources; AHP, on the other hand, is a strong way to solve complex problems. Integrating of GIS and AHP methods provides decision-makers with a perfect and immediate review at the lowest cost. In this example, the decision-making process began by examining 12 criteria and obtaining available information based on the criteria for each layer and combined them, then by designing a questionnaire and used the experts opinion to weight the criteria with the AHP method and finally with the help of These weighted layers which are combined in GIS software.

The present study showed that very suitable landfills for the waste of Paveh county cover an area of 51.68794 km^2 or about 7% of the total classified sites.

Acknowledgements

I would like to thank dear editor Ivana Barac, who patiently followed the work steps on my behalf, and Professor Fabio De Felice, who gave me this opportunity.

Author details

Mahya Rostampoor
Faculty of Engineering, Department of Civil Engineering, University of Urmia, Urmia, Iran

*Address all correspondence to: mahyarostampoor@gmail.com

References

[1] Saaty TL. Decision making with the analytic hierarchy process. International Journal of Services Sciences. 2008;**1**(1): 83-98

[2] Bhushan N, Rai K. Strategic Decision Making: Applying the Analytic Hierarchy Process. Springer Science & Business Media; 2007

[3] Khaira A, Dwivedi R. A state of the art review of analytical hierarchy process. Materials Today: Proceedings. 2018;**5**(2):4029-4035

[4] Janssen M, van der Voort H, Wahyudi A. Factors influencing big data decision-making quality. Journal of Business Research. 2017;**70**:338-345

[5] Albayrak E, Erensal YC. Using analytic hierarchy process (AHP) to improve human performance: An application of multiple criteria decision making problem. Journal of Intelligent Manufacturing. 2004;**15**(4): 491-503

[6] Forman EH, Selly MA. Decision by Objectives: How to Convince Others that you Are Right. World Scientific; 2001

[7] Eastman JR, Jiang H, Toledano J. Multi-criteria and multi-objective decision making for land allocation using GIS. In: Multicriteria Analysis for Land-Use Management. Springer; 1998. pp. 227-251

[8] Saaty TL, Bennett J. A theory of analytical hierarchies applied to political candidacy. Behavioral Science. 1977; **22**(4):237-245

[9] Rostampoor M, Zamani M, Vaighan A. Combining GIS and analytical hierarchy process for landfill siting, study area: Paveh County in Iran. Journal of Civil Engineering Frontiers. 2020;**1**(1):07-15

[10] Kazakidis V, Mayer Z, Scoble M. Decision making using the analytic hierarchy process in mining engineering. Mining Technology. 2004;**113**(1):30-42

[11] Ramanathan R. A note on the use of the analytic hierarchy process for environmental impact assessment. Journal of Environmental Management. 2001;**63**(1):27-35

[12] Danesh D, Ryan M, Abbasi A. Using analytic hierarchy process as a decision-making tool in project portfolio management. WASET International Journal of Economics and Management Engineering. 2015;**9**(12):4194-4204

[13] Bascetin A. A decision support system using analytical hierarchy process (AHP) for the optimal environmental reclamation of an open-pit mine. Environmental Geology. 2007; **52**(4):663-672

[14] Saaty TL. Decision Making for Leaders: The Analytic Hierarchy Process for Decisions in a Complex World. RWS Publications; 2001

[15] Harker PT. The art and science of decision making: The analytic hierarchy process. In: The Analytic Hierarchy Process. Springer; 1989. pp. 3-36

[16] Nagesha N, Balachandra P. Barriers to energy efficiency in small industry clusters: Multi-criteria-based prioritization using the analytic hierarchy process. Energy. 2006;**31**(12): 1969-1983

[17] Saaty RW. The analytic hierarchy process—What it is and how it is used.

Mathematical Modelling. 1987;**9**(3–5): 161-176

[18] Saaty TL. What is the analytic hierarchy process? In: Mathematical Models for Decision Support. Springer; 1988. pp. 109-121

[19] Sadeghpour M, Mohammadi M. Evaluating traffic risk indexes in Iran's rural roads. Case study: Ardabil-Meshkin rural road. Transport and Telecommunication. 2018;**19**(2):103

[20] Stein WE, Mizzi PJ. The harmonic consistency index for the analytic hierarchy process. European Journal of Operational Research. 2007;**177**(1): 488-497

[21] Lin Z-C, Yang C-B. Evaluation of machine selection by the AHP method. Journal of Materials Processing Technology. 1996;**57**(3–4):253-258

[22] Sinuany-Stern Z, Mehrez A, Hadad Y. An AHP/DEA methodology for ranking decision making units. International Transactions in Operational Research. 2000;**7**(2): 109-124

[23] Jankowski P. Integrating geographical information systems and multiple criteria decision-making methods. International Journal of Geographical Information Systems. 1995; **9**(3):251-273

[24] Mikhailov L, Singh MG. Comparison analysis of methods for deriving priorities in the analytic hierarchy process. In: IEEE SMC'99 Conference Proceedings. 1999 IEEE International Conference on Systems, Man, and Cybernetics (Cat. No. 99CH37028). IEEE; 1999

[25] Saaty TL, Vargas LG. Comparison of eigenvalue, logarithmic least squares and least squares methods in estimating ratios. Mathematical Modelling. 1984; **5**(5):309-324

[26] Calantone RJ, Di Benedetto CA, Schmidt JB. Using the analytic hierarchy process in new product screening. Journal of Product Innovation Management. 1999;**16**(1):65-76

[27] Saaty TL, Vargas LG. Prediction, Projection and Forecasting: Applications of the Analytic Hierarchy Process in Economics, Finance, Politics, Games and Sports. Springer; 1991

[28] Wagner ED. Public Key Infrastructure (PKI) and Virtual Private Network (VPN) Compared Using an Utility Function and the Analytic Hierarchy Process (AHP). Virginia Tech; 2002

[29] Marinoni O. Implementation of the analytical hierarchy process with VBA in ArcGIS. Computers & Geosciences. 2004;**30**(6):637-646

[30] Saaty TL, Vargas LG. The decision by the US congress on China's trade status: A multicriteria analysis. In: Models, Methods, Concepts & Applications of the Analytic Hierarchy Process. Springer; 2001. pp. 305-317

Chapter 10

Decision to Use Electric Vehicle at Airport Operation in Jakarta

Mustika Sari and Reni Dian Octaviani

Abstract

The purpose of this chapter is to determine the options and solutions used in the choice to deploy electric ground support equipment at Jakarta's Soekarno-Hatta airport in order to reduce emissions. The problem in this study is that there are significant emissions at Soekarno-Hatta airport which cause pollution in the Soekarno-Hatta airport environment. The research method used an analytical networking process (ANP). Respondents used were 15 people consisting of experts in the fields of the airline, ground handling, and airport authority. The interview techniques used were in-depth interviews. The results of interviews with experts in the field of airlines, ground handling, and airports said that the use of ground power units (GPUs), the use of low-emission cars, energy saving controls, and the installation of airport surface management systems are the four problem criteria. Utilizing low-emission automobiles, which have a weight value of 0.3074, is the problem's primary criterion out of the four components. Meanwhile, there are three alternative solutions, namely Environmental Education, Environmental Innovations, and Aircraft carbon management. Of the three alternatives, Aircraft carbon management is the main alternative solution with a weight value of 0.3530.

Keywords: electric vehicle, airport operation, analytical networking process, government regulations, aircraft carbon management, Jakarta

1. Introduction

Aircraft serve as a means of transportation for passengers and cargo, offering unique benefits such as speed and cost efficiency that are not available with other modes of transportation, on the other hand, air transportation also produces exhaust gases which are pollutants that damage the ozone layer. Although it only accounts for 3% of total air emissions, the number of aircraft will increase the number of emissions in the coming years [1].

The main function of an airport is to provide aviation and non-aviation-related services to users, namely airlines, passengers, and cargo. Airports also provide ground handling. But some airlines have their own ground handling, and there are also companies specialized in handling airplanes while on the ground or called ground-handling companies [2].

Ground handling or aircraft handling while at the airport includes a fairly long process, which starts with handling passengers to report themselves (check-in) until the passengers enter the aircraft and the door is closed. In addition to passengers, this handling is also for luggage, cargo, and the aircraft itself. Such as cleaning the aircraft cabin, and pushing the plane to park baggage handling for cargo and passenger luggage [3].

Operational handling of aircraft handling is needed when the aircraft is on the ground, such as ramp handling, and on-board services, but depending on the aircraft's operational budget [4].

Expedite the task necessitated the use of some additional equipment for the transportation of planes, people, and cargo while on the ground. The equipment that is prepared to fulfill the demands of the aircraft while the aircraft is on the ground, both during the departures, arrivals, and transits. The equipment being used is called Ground Support Equipment [4].

Most GSE is usually associated with aircraft servicing while the aircraft is on the ground. The activity starts when the block is on until the block is off. Until the block is lifted and the aircraft is ready for takeoff, chocks are placed in front of the wheels. While on the ground, ground-handling tasks include loading, and unloading passengers and baggage, aircraft cleaning and maintenance, refueling, and other services [5].

One of the high sources of pollution at airports is ground vehicle activity. By 2050, carbon dioxide emissions will be produced by the aviation sector. And it will grow up to 2 to 10 times as it is compared to emissions in the early 2000s [6].

Currently, ground support equipment emission reduction measures are numerous and varied at the airports. Initiatives are also coming from airport operators and GSE providers. There are several examples of GSE emission reduction measures implemented over the past few years. Many airports around the world have emission reduction programs and alternative fuels [5].

Airport authorities, in response to community concerns, public awareness of environmental issues from aviation activities, and regulatory measures of governments and local authorities, have tried to implement strategies and procedures to reduce the adverse impact of their activities on the environment, appropriately and effectively.

Indonesia's Directorate General of Civil Aviation (DGCA), and Ministry of Transportation, submitted to ICAO in 2015 a Country Action Plan to diminish greenhouse gas emissions in the aviation area which will be updated in 2018 as demanded by members. This shows Indonesia's commitment to supporting global policies on aviation environmental protection (Indonesia [7]). The problems in this study are how is the government policy on the use of electric and diesel for GSE equipment and what is the conceptual model for decision-making to use GSE equipment with electric fuel?

The success of airline corporations in achieving on-time performance, flight safety, and customer happiness is greatly aided by ground-handling businesses. Ground handling is responsible for handling passengers and aircraft at the airport as well as during takeoff and landing. Passengers, passenger bags, cargo, postal items, and ramp handling are among items that ground employees manage. The range of work involved in ground handling is regulated in the IATA Airport Handling Manual [8] which consists of 9 standard service sections namely; the Passengers Handling, Baggage Handling, Mail and Cargo Handling, Aircraft Handling and Loading, Load Control, Air Side Management, and Safety, Aircraft Movement Control, Standard Ground-Handling Agreement, and the Airport Handling GSE Specification.

Among all of services is the Airport Handling GSE Specification which specifically handles aircraft handling while on the ground with the backup of ground support equipment depends on the type of airplanes. To accelerate the work, some auxiliary equipment is needed for the movement of aircraft, passengers, and cargo while on the ground. Equipment that is prepared to support the aircraft needs while the aircraft is on the ground, both at the time of departure and arrival or transit, the equipment is known as GSE (Ground Support Equipment). GSE equipment is classified into two parts based on its movement ability or workability, namely Motorized and Non-Motorized GSE [9].

Handling activities associated with aircraft during ground time contribute significantly to air pollution at airports. The activities include all machinery and vehicles and that service aircraft in their parked position (e.g., high loaders, conveyor belts, passenger ladders) and those at airside and ramp areas (e.g., toilet trucks, cargo tractors, catering trucks).

Emissions from such sources must be evaluated for a variety of pollutants in order to effectively manage air quality. Goal setting, technology observation, and mitigation planning can all benefit from this information. Information on GSE vehicles can be obtained from various handling agents and lessors. We can also find out the GSE's age, engine size and type, and total operating time per year and/or fuel consumption. Further evaluation reveals that the limits of the emission are usually not used by manufacturers of non-road mobile machinery and that certain advantages are being maintained. Emissions from a given GSE are based on aircraft movements and correspond to aircraft type [10].

Pollution generated at airports results in potentially harmful levels of pollution. Exhaust from aircraft and diesel engines, direct fuel emissions from aircraft refueling, and larger dust particles from brakes, tires, asphalt, dirt, and so on are the primary sources of air pollution at airports. Polycyclic aromatic hydrocarbons (PAHs), volatile organic compounds (VOCs), inorganic gases like sulfur dioxide (SO2) and nitrogen oxides (NOx), and particulate matter (PM) are all examples of pollutants [11].

Pollution generated from aircraft harms the environment. Although it only accounts for 3% of the world's total emissions, the number of aircraft in operation is expected to increase. The aviation industry and aviation business services contribute to the increase in greenhouse emissions. The contribution of emissions from the aviation industry has shared as one of the five largest contributors to the largest emissions, which is below 5% [1].

In a previous study, it was said that the aviation sector has contributed 12% of carbon emissions. The study also estimates that by 2050 carbon dioxide (CO2) emissions which are produced by the aviation area will grow up to 2 to 10 times when compared to emissions in the early 2000s [6].

Governments and International Aviation (IATA) have committed to reducing pollution in aviation by building a four-pillar strategy, namely improved aviation technology, effective flight operations, efficient infrastructure development, and economic measurement of carbon emission markets. In 2011 Luthfansa conducted a flight experiment with biofuel on an Airbus A321 aircraft and successfully executed 8 hours and 20 minutes of flight time, and claimed to be able to reduce carbon dioxide (CO2) emissions by 38 tons.

Meanwhile, research from Oxford University states that there are other ways to reduce aviation emissions by 95%, including direct flights and not waiting long to land. And the best time on the apron and or runway is not too long after the engine is running.

Government and stakeholder support is needed in this case airlines, airport operators, and ground handling. Ground handling plays a role in efforts to reduce emissions, especially related to the ground support equipment (GSE) operation at the airport. This is carried out by planning and making gradual implementation of vehicles and ground services tools (GSE) based on electricity or renewable fuel (biofuel). Other aspects that have a significant role in reducing aviation emissions are the airport operators. Greening around the airport environment by planting trees to absorb carbon dioxide (CO2) exhausts emissions on access roads to airport terminals, parking areas, and other open areas. Tree planting is done by planting trees such as tamarind which absorbs about 28 tons of CO2/per tree/per year [6].

Based on the results of the discussion and description [12] Soekarno-Hatta International Airport is vulnerable to many problems. Problems accumulate, ranging from operational to non-operational problems. If this is not resolved immediately, the accumulated problems will become more complicated with great risk. These conditions will increase the risk of noise factors, and the potential for health problems for people living and doing activities in the airport area. The results show that the noise implications of the model can be constructed through the analysis of factors that can provide an overview of the effect of noise on society. This research concludes that modeling environmental capacity produces several noise indicators that can be used as instruments to evaluate, build, and develop airport noise control policies as part of controlling eco-airport performance.

Transportation accounts for global CO2 emissions as much as 26%. Moreover, it is one of the few industrial sectors where emissions are still increasing. The main contributors to greenhouse gas emissions from the transportation sector are automobile use, road transport, and aviation. Although new technologies, particularly those that have the potential to replace the use of petroleum in transportation, are discussed, it appears that technology alone cannot address climate change. Changes in behavior through policy are also essential for stabilizing transportation emissions of greenhouse gases. Policymakers are facing increasing pressure to address climate change in transportation. Even though it is common to focus on future technological solutions, in order to fully benefit from new technologies, it is important to make immediate behavioral changes [13].

When viewed from previous studies, the use of fuel in the aviation industry can cause a decrease in air quality as well as the ground support industry. Meanwhile, Haryato's journal mentions that in addition to fuel that causes pollution, aircraft noise also causes noise pollution. In a journal written by Lee Chapman that transportation accounts for 26% of carbon dioxide (CO2) emissions as cars, cargo transportation, and the aviation industry. While research on the use of biofuels in ground-handling fuel has never been done, for this reason, researchers try to conduct research on the use of biofuel in ground-handling equipment for movement on the apron.

This chapter is divided into four parts. The first part talks about the conceptual model for decision-making to use GSE equipment with electric fuel and government policy on the use of electric and diesel for GSE equipment. In this part there is a discussion that investigates the emissions from airport ground-handling activities and their impact on air quality. We will then go over the various measures that have been put in place to reduce emissions from GSE equipment, as well as their effectiveness. The second part is about the research method used to study the impact of electric and diesel Ground Service Equipment (GSE) on ground-handling companies. The third part is about the result and discussion. Finally, there is a summary of the findings and suggestions for future research in this area.

2. Method

This study utilizes primary data collected through in-depth interviews with experts and industry professionals to gain a thorough understanding of the issue at hand and gather information from ground-handling companies. The process includes first bringing attention to the problem, then conducting interviews with experts and practitioners, and using a paired questionnaire to gather further data [14].

This research also employs secondary data sourced from outside sources, specifically data from ground-handling companies and providers of GSE equipment. The data includes the cost of electric GSE and diesel GSE.

To gather this information, researchers establish protocols or forms for recording data and devise methods for collecting it, such as interviews or observation techniques [15].

The research population for preliminary interviews was 15 people from ground-handling companies, airlines, and airport authorities.

The study included 15 people from ground-handling company, airlines, and airport authorities as followed:

a. Ground-Handling company (Senior Manager, Operation Manager, General Manager, Senior Operator)

b. PT Garuda Indonesia (Pilot, Ex pilot, Operation Manager)

c. Airport Authority (Operational Manager Airport, Marshaling, and Senior Operator)

The decision model in this study was developed using a combination of primary and secondary data collection methods. Primary data was gathered through in-depth interviews with industry experts and professionals, while secondary data was obtained from literature review as a guidance for the indicator and interview with the experts.

Data analysis techniques using Analytical Networking Process (ANP), with Super Decision tools. The ANP (Analytic Network Process) is a mathematical approach that can evaluate the impact of different strategies and assumptions on solving a problem. This method is applied by adjusting the complexity of the problem and the priority scale that produces the greatest priority effect [16].

In contrast to the conventional mathematical models that economics typically use for quantitative analysis, the ANP method offers economists a unique approach to addressing economic issues. The ANP approach is based on a more robust absolute scale that is used to represent pairwise comparisons of valuations.

The foundation of homogeneity in ANP states that in the ANP framework, it is important that the elements being compared are similar to one another, as significant differences between them can result in larger inaccuracies in determining the value of the factors that influence the decision.

Table 1 defines the numerical rating scale to be used when conducting interviews. The scale is defined with values from 1 to 9.

In order to find the answers to the research questions while achieving research objectives, the research stages will be as follows:

Intensity of Importance	Definitions
1	Equal importance
2	Weak
3	Moderate importance
4	Moderate plus
5	Strong importance
6	Strong plus
7	Very strong
8	Very very strong
9	Extreme importance
Source: [16].	

Table 1.
Definition of rating scale and numeric scale.

a. Making model construction, which can be obtained from the results of interviews, and literature review.

b. Quantification of the model, namely the preparation of a questionnaire based on the results of interviews with experts.

c. Analysis of results, namely the results of the questionnaire are processed to get the interpretation of the results.

3. Result and discussion

3.1 Government regulations on the use of electrical fuel and diesel fuel in GSE vehicles at airports in Jakarta

Presidential Regulation No.55 of 2019 is the Fast Track Program for Battery Electric Vehicles for Road Transport (Perpres No. 55, 2019). This presidential regulation aims:

a. to promote energy efficiency, security, and conservation in the transportation sector, as well as clean air quality and a reduction of greenhouse gas emissions, it is important to accelerate the implementation of battery electric vehicles for road transportation in Indonesia.

b. to provide guidance, a foundation, and legal certainty for the acceleration of the road transportation battery electric vehicle program.

c. to foster proficiency in industrial technology and vehicle design, and establish Indonesia as a hub for the manufacturing and export of motor vehicles, it is essential to accelerate the battery electric vehicle program [17].

Supported by the DKI Jakarta Governor's regulation no. 3 of 2020 concerning Motor Vehicle Transfer Tax Incentives for Battery Electric Vehicles for Road Transportation, that:

a. Battery Electric Vehicles hereinafter referred to as Battery-Based KBL, are vehicles that are driven by Electric Motors and get a supply of electric power resources from batteries directly in the vehicle or from outside.

b. Motor Vehicle Title Transfer Fee is a tax on the transfer of ownership rights of motor vehicles as a result of an agreement between two parties or unilateral actions or circumstances that occur due to sale and purchase, exchange, grants, inheritance, or entry into a business entity.

c. Motor Vehicle Tax and Motor Vehicle Title Transfer Tax Service Unit, hereinafter referred to as UP PKB and BBN-KB, is a Technical Implementation Unit of the Regional Tax and Retribution Agency of the Special Capital Region of Jakarta Province that carries out Motor Vehicle Tax and Motor Vehicle Title Transfer Tax collection services [18].

Electric vehicles are seen as the greatest potential way to decrease air pollution, as they are considered to be zero-emission vehicles (ZEVs), due to the absence of a fossil-fueled internal combustion engine (IC) in the powertrain structure. However, due to some technical specifications and structural design (the presence of a battery driving the electric motor), currently, EV batteries and rechargeable accumulators are charged by external electric power sources. Thus, although EVs are considered ZEV vehicles, they can still pollute the environment indirectly, due to the emission of pollutants coming from the electricity production process at the power plant [19].

The integration of electric vehicles in air transportation can be seen in the use of e-GSE, or electric ground support equipment, which assists with aircraft operations on the ground. This includes the use of electric vehicles for ground support tasks at airports (**Table 2**).

Create a conceptual model for decision-making in the use of electric fuel for ground support equipment using the analytical networking process method.

No	GSE Vehicle	Abbreviation
1	Pax Boarding Stair	e-PBS
2	Belt Conveyor Loader	e-BCL
3	Baggage Towing Tractor	e-BTT
4	High Lift Loader	e-HLL
5	Forklift	e-FTL
6	Push Back/ATT Wide Body	e-ATT
7	Ground Power Unit 180 KVA	e-GPU

Source: [20].

Table 2.
The electric ground support equipment.

3.1.1 Criteria and sub-criteria

To make it easier for respondents to understand each dimension and indicator, or in this method called criteria and sub-criteria, the following is an explanation of each dimension.

3.1.1.1 Energy conservation and carbon reduction in airport operation

Implementing an airport surface management system (SMAN) that optimizes taxi lanes, reducing taxi, and wait times, which in turn reduces carbon emissions.

3.1.1.2 Use of renewable resources

Utilizing renewable energy sources such as solar, wind, and biomass to power airport operations.

3.1.1.3 Airport environmental sustainability management

Regular monitoring of air quality and holding airlines accountable for their emissions (**Table 3**) [16].

Dimension	Indicator	Definition
Energy conservation and carbon reduction in airport operations	a. Installation of airport surface management system	A surface management system (SMAN) is implemented in airports to decrease carbon emissions by directing aircraft to the optimal taxi routes, thus reducing the time spent taxing and waiting in queue.
	b. Use of ground power units (GPUs)	Maximizing the use of ground power units (GPUs) in place of aircraft auxiliary power units (APUs) is implemented.
	c. Use of low-emission vehicles	Maximizing the use of vehicles that produce low emissions and are powered by natural gas, electricity, and hybrid-electric power is implemented.
	d. Installation of water-saving devices	Water conservation is achieved by installing devices such as water-efficient toilets, water economizers, and low-flow taps.
	e. Energy-saving control	The energy consumption of facilities such as lighting, heating, and cooling systems, escalators, and moving walkways is managed efficiently to reduce energy use.
Use of renewable resources	a. Use of renewable energies	Maximizing the utilization of renewable energy sources like solar, wind, and biomass is implemented.
	b. Use of recycled water	The recycling and reuse of wastewater and rainwater for purposes such as flushing toilets, watering plants, and cleaning is implemented.

Dimension	Indicator	Definition
	c. Waste recycling and reuse	Implementing measures to decrease the amount of waste produced and recycling and reusing waste is implemented.
	d. Recycling of kitchen waste and wastewater	Airport eateries are mandated to recycle kitchen waste and wastewater.
Airport environmental sustainability management	a. Aircraft carbon management	Regular monitoring of air quality around airports is done through the use of air-quality monitoring stations, and airlines are held accountable for air pollution through charges.
	b. Aircraft noise management	Regular monitoring of noise around airports is carried out through noise monitoring stations and airlines are held accountable for noise pollution through charges.
	c. Environmental education and its effectiveness	Airports promote and support energy conservation, carbon reduction, and environmental protection among all its affiliates and evaluate the success of these efforts.
	d. Environmental and ecological conservation	Airports are dedicated to improving the environment and preserving ecology through efforts such as beautifying their surroundings.
	e. Green features and environmental innovations	Airports are proud to showcase unique and innovative solutions for energy conservation, carbon reduction, and environmental protection that they have implemented.

Source: [21].

Table 3.
Criteria and sub-criteria.

3.1.2 Research process

To gain insight into the decision to use electric fuel for ground support equipment, research stages will be conducted following the research plan outlined in the research chart as follows:

To find out the most prioritized criteria, sub-criteria, and strategies in the decision model for selecting the use of electric fuel in ground support equipment, there are three phases or stages of research that will be carried out. The three phases are as follows:

3.1.2.1 Model construction

The ANP model is built by conducting theoretical and empirical literature reviews, consulting experts and practitioners through questionnaires or interviews, and conducting in-depth interviews to gain a deeper understanding of the issues and problems faced in the field.

3.1.2.2 Model quantification

The quantification stage of the model involves using questions in the ANP survey in the form of pairwise comparisons between elements in a cluster to determine which element has a stronger influence and the degree of that influence, using a numerical scale of 1–9. After that the assessment data is collected and inputted through expert choice 11.0 software to be processed to produce output in the form of priorities and matrices. The results of each respondent will be inputted into a separate ANP network.

3.1.2.3 Synthesis and analysis

In synthesizing and analyzing the geometric mean instrument is used. Geometric Mean is used to determine the results of individual assessments from respondents and determine the results of opinions in a group [16]. Questions in the form of pairwise comparisons from respondents will be combined to form a consensus. Geometric mean is a method of calculating the average that indicates a particular tendency or value. It is calculated using the following formula:

$$\left(\prod_{i}^{n} = 1a_i\right)^{1/n} = \sqrt[n]{a_1 a_2, a_n} \quad (1)$$

n = Respondent 1. … … .. n.
i = Pairwise 1 … … ..i.

3.1.3 Data collection process

Data collection is carried out using interviews and assisted by questionnaires. During interviews with experts and practitioners, a pairwise comparison method will be employed to determine the weight of each indicator. This is achieved by giving an open-ended questionnaire to the interviewees.

Expert Choice software version 11.0 will be used to conduct pairwise comparisons. This software is commonly used in ANP research. An example of a questionnaire in conducting pairwise comparisons is shown in **Figure 1.**

3.2 Interview results with experts from airlines, ground handling, and airport authority

3.2.1 Results of interviews with experts from airlines

The following are the results of interviews with experts from airlines, where the average opinion is that the most important criterion in supporting using electric GSE is the use of low-emission vehicles with a value of 0.30 and the highest alternative or solution is environmental innovations with a value of 0.36 (**Table 4**).

3.2.2 Results of interviews with experts from ground handling

The following are the results of interviews with experts from ground handling, where the average opinion is that the most important criteria in supporting the use of electric GSE are the use of low-emission vehicles with a value of 0.32 and the highest alternative or solution is aircraft carbon management with a value of 0.38 (**Table 5**).

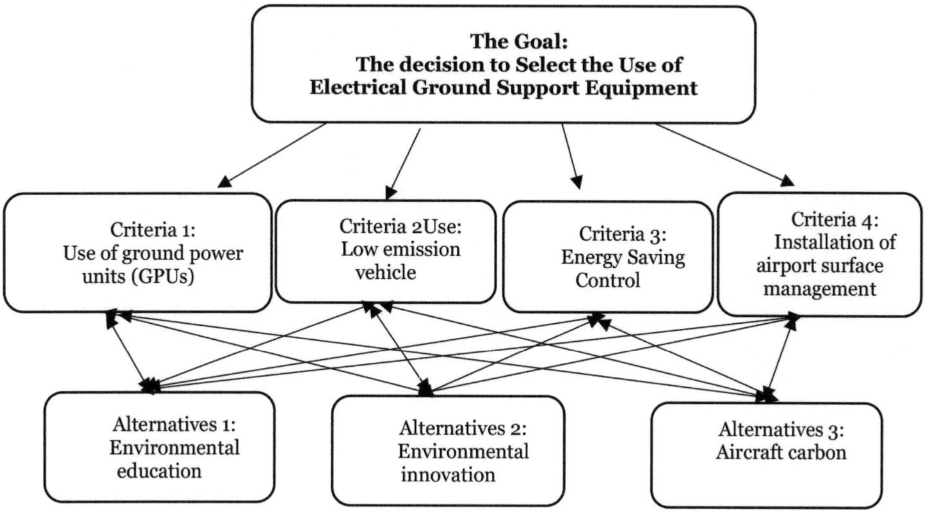

Figure 1.
Analytic network process structure model.

Element	Normalized by Cluster					Mean
Alternative	**R1**	**R2**	**R3**	**R4**	**R5**	
Aircraft Carbon Management	0,34,696	0,33,245	0,38,366	0,34,034	0,24,115	0,32
Environmental Education	0,33,323	0,26,419	0,26,472	0,33,434	0,36,781	0,31
Environmental Innovations	0,31,981	0,40,337	0,35,161	0,32,532	0,39,104	0,36
Criteria						
Energy Saving Control	0,18,415	0,28,848	0,24,141	0,2628	0,20,316	0,24
Installation of Airport Surface Management System	0,27,869	0,24,543	0,30,201	0,22,199	0,32,188	0,27
Use of Low-Emission Vehicle	0,42,763	0,20,387	0,34,403	0,23,803	0,30,846	0,30
Use of Ground Power Units (GPUs)	0,10,953	0,26,222	0,11,255	0,27,717	0,1665	0,19

Table 4.
Interview results with airline experts.

3.2.3 Results of interviews with experts from the airport authority

The following are the results of interviews with experts from the airport authority, where the average opinion is that the most important criteria in supporting using electric GSE are the use of low-emission vehicles with a value of 0.32 and the highest alternative or solution is aircraft carbon management with a value of 0.36 (**Table 6**).

3.3 The outcome of the interviews conducted with specialists in the airline, ground handling, and airport authority industries

Based on the diagram below, there are four problem criteria, namely, the use of ground power units (GPUs), the Use of low-emission vehicles, Energy Saving Control,

Element	Normalized by Cluster					Mean
Alternative	**R6**	**R7**	**R8**	**R9**	**R10**	
Aircraft Carbon Management	0,38,761	0,34,696	0,43,414	0,38,366	0,35,558	0,38
Environmental Education	0,25,976	0,33,323	0,23,601	0,26,472	0,32,389	0,28
Environmental Innovations	0,35,263	0,31,981	0,32,985	0,35,161	0,32,053	0,33
	Criteria					
Energy Saving Control	0,18,831	0,18,415	0,27,898	0,24,141	0,25,959	0,23
Installation of Airport Surface Management System	0,40,405	0,27,869	0,21,408	0,30,201	0,22,766	0,29
Use of Low-Emission Vehicle	0,31,733	0,42,763	0,21,579	0,34,403	0,27,718	0,32
Use of Ground Power Units(GPUs)	0,09031	0,10,953	0,29,116	0,11,255	0,23,557	0,17

Table 5.
Interview results with ground-handling experts.

Element	Normalized by Cluster					Mean
Alternative	**R11**	**R12**	**R13**	**R14**	**R15**	
Aircraft Carbon Management	0,35,216	0,38,808	0,34,737	0,38,476	0,31,178	0,36
Environmental Education	0,30,886	0,25,938	0,31,212	0,29,136	0,28,026	0,29
Environmental Innovations	0,33,898	0,35,253	0,34,051	0,32,388	0,40,796	0,35
	Criteria					
Energy Saving Control	0,19,625	0,19,392	0,19,749	0,2496	0,29,396	0,23
Installation of Airport Surface Management System	0,34,124	0,40,281	0,33,987	0,29,738	0,09543	0,30
Use of Low-Emission Vehicle	0,30,546	0,31,579	0,2915	0,33,876	0,35,474	0,32
Use of Ground Power Units(GPUs)	0,15,704	0,08748	0,17,115	0,11,427	0,25,587	0,16

Table 6.
Interview results with airport authority expert.

and the Installation of an airport surface management system. Of the four aspects, the Use of low-emission vehicles is the main criterion in the problem with a weight value of 0.3074. The Installation of airport surface management system aspect is the second problem criterion with a weight value of 0.2716, then the Energy Saving Control aspect is the third problem criterion with a weight value of 0.2276 and there is the Use of ground power units (GPUs) aspect which is the last priority in the problem criteria weights 0.1559. The rater agreement value of this criterion is 0.4933 (w = 0.4933) which is on a moderate to strong scale, indicating that experts agree in determining the problem criteria in the study (**Figures 2** and **3**).

From the diagram above, there are three alternative solutions, namely Environmental Education, Environmental Innovations, and Aircraft carbon management. Of the three alternatives, Aircraft carbon management is the main alternative solution

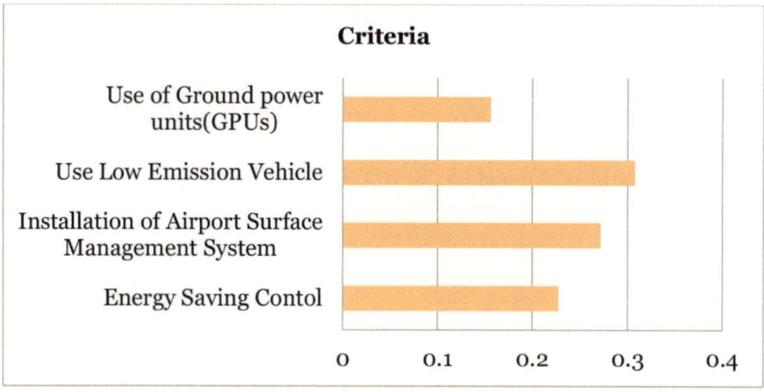

Figure 2.
Aspects of criteria (interview result).

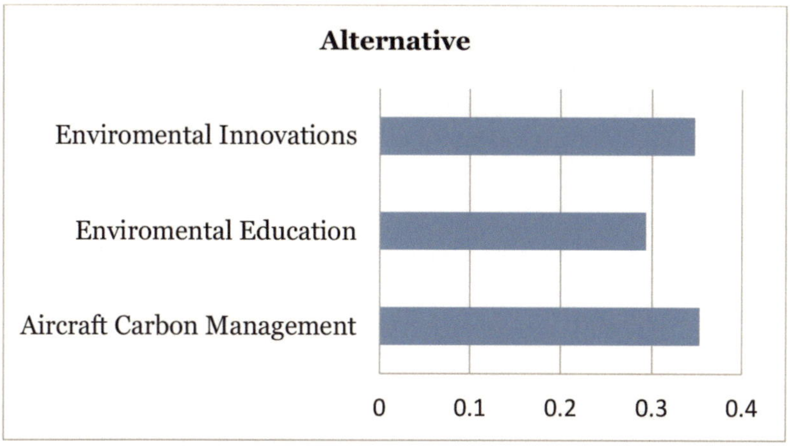

Figure 3.
Alternative aspects (interview result).

with a weight value of 0.3530. Furthermore, Environmental Innovations is the second alternative solution with a weight value of 0.3474, and Environmental Education is the last priority in alternative solutions with a weight of 0.2933. The rater agreement value of this criterion is 0.2497 (w = 0.2497) which is on a weak to moderate scale, indicating that the experts' answers vary in determining alternative solutions.

One of the high sources of pollution at the airport is ground vehicle activity. Emissions caused by the Aircraft Power Unit (APU) are quite high ranging from 2071 to 2892 kg/hr., quite high according to emission standards.

The results of interviews with experts in the fields of airlines, ground handling, and airports said that of the four problem criteria, namely Use of ground power units (GPUs), Use of low-emission vehicles, Energy Saving Control, and Installation of airport surface management systems. Of the four aspects, the Use of low-emission vehicles is the main criterion in the problem with a weight value of 0.3074.

There are three alternative solutions, namely Environmental Education, Environmental Innovations, and Aircraft carbon management. Of the three alternatives,

Aircraft carbon management is the main alternative solution with a weight value of 0.3530.

A large number of activities and a variety of measures to reduce ground support equipment emissions at airports have been carried out. Initiatives also come from airport operators and ground support equipment (GSE) providers.

Indonesia's Directorate General of Civil Aviation (DGCA), and Ministry of Transportation, submitted to ICAO in 2015 a Country Action Plan to reduce greenhouse gas emissions in the aviation sector which will be updated in 2018 as requested by members. This demonstrates Indonesia's commitment to supporting global policies on aviation environmental protection.

The Analytic Hierarchy Process (AHP) and Analytic Network Process (ANP) are multi-criteria decision-making (MCDM) methods that can be used to analyze complex problems with multiple criteria and interdependencies. They are often used in transportation studies to evaluate different options, such as the feasibility of using electric ground support equipment. In this study, the ANP method was used to develop a conceptual model for the implementation of electric fuel for ground support equipment by interviewing experts in the airline, ground handling, and airport authority industries. The ANP method is considered appropriate for this subject because it allows for the evaluation of multiple criteria and the relationships between them, providing a comprehensive understanding of the problem and potential solutions [22].

4. Conclusion

There are three alternative solutions, namely Environmental Education, Environmental Innovations and Aircraft carbon management. Of the three alternatives, Aircraft carbon management is the main alternative solution with a weight value of 0.3530. A large number of activities and a variety of measures to reduce ground support equipment emissions at airports have been carried out. Initiatives also come from airport operators and ground support equipment (GSE) providers.

Indonesia's Directorate General of Civil Aviation (DGCA), Ministry of Transportation, submitted to ICAO in 2015 a Country Action Plan to reduce green house gas emissions in the aviation sector and will be updated in 2018 as requested by members. This demonstrates Indonesia's commitment to supporting global policies on aviation environmental protection.

A. Appendix

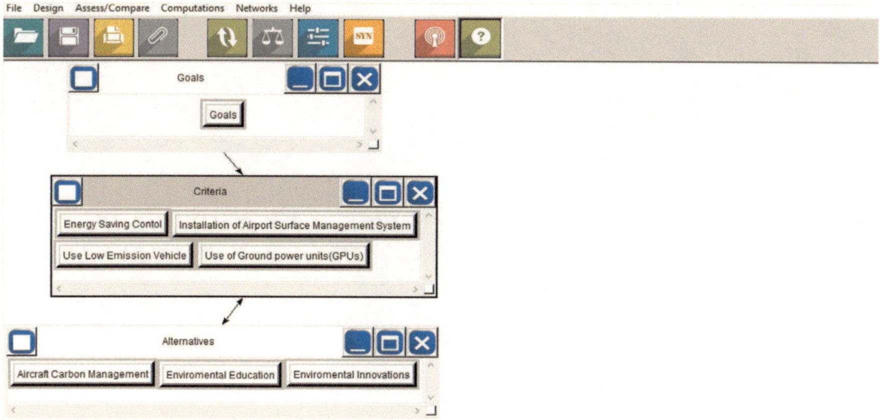
File Design Assess/Compare Computations Networks Help
Goals
Goals
Criteria
Energy Saving Contol
Installation of Airport Surface Management System
Use Low Emission Vehicle
Use of Ground power units(GPUs)
Alternatives
Aircraft Carbon Management
Enviromental Education
Enviromental Innovations

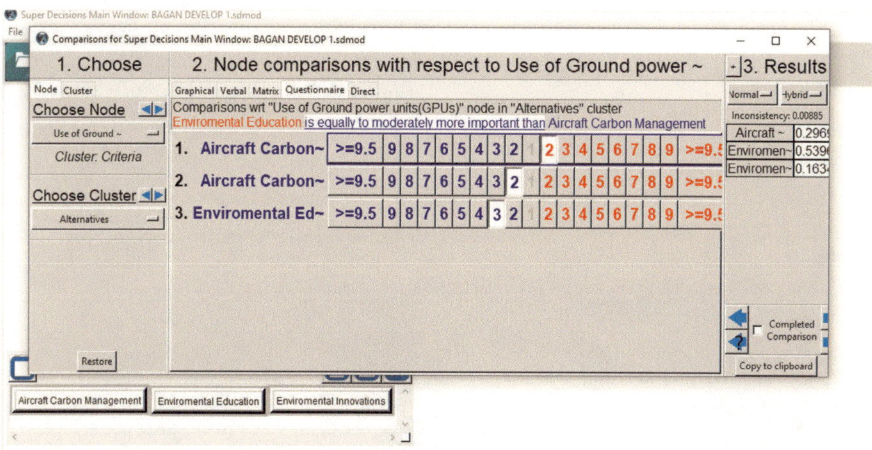
Comparisons for Super Decisions Main Window: BAGAN DEVELOP 1.sdmod
1. Choose
2. Node comparisons with respect to Use of Ground power ~
3. Results
Choose Node
Use of Ground ~
Cluster: Criteria
Choose Cluster
Alternatives
Comparisons wrt "Use of Ground power units(GPUs)" node in "Alternatives" cluster
Enviromental Education is equally to moderately more important than Aircraft Carbon Management
1. Aircraft Carbon~
2. Aircraft Carbon~
3. Enviromental Ed~
Inconsistency: 0.00885
Completed Comparison
Copy to clipboard
Restore
Aircraft Carbon Management
Enviromental Education
Enviromental Innovations

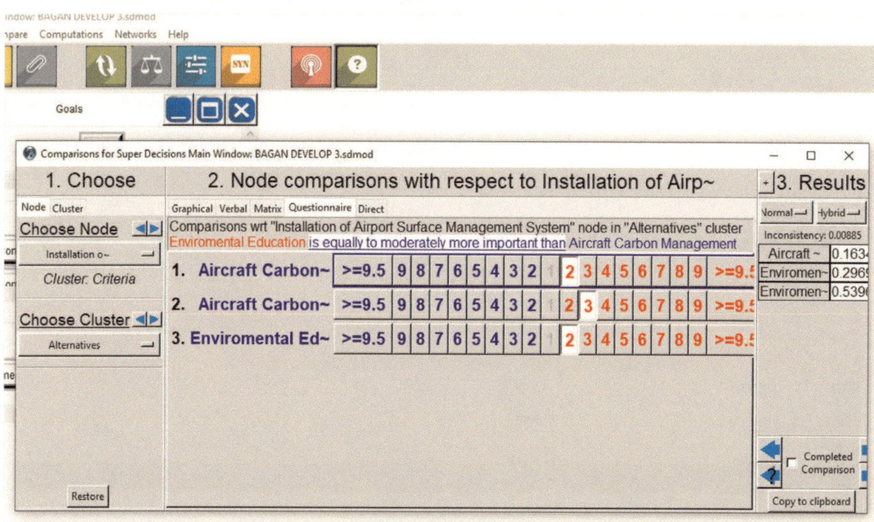
Comparisons for Super Decisions Main Window: BAGAN DEVELOP 3.sdmod
1. Choose
2. Node comparisons with respect to Installation of Airp~
3. Results
Choose Node
Installation o~
Cluster: Criteria
Choose Cluster
Alternatives
Comparisons wrt "Installation of Airport Surface Management System" node in "Alternatives" cluster
Enviromental Education is equally to moderately more important than Aircraft Carbon Management
1. Aircraft Carbon~
2. Aircraft Carbon~
3. Enviromental Ed~
Inconsistency: 0.00885
Completed Comparison
Copy to clipboard
Restore

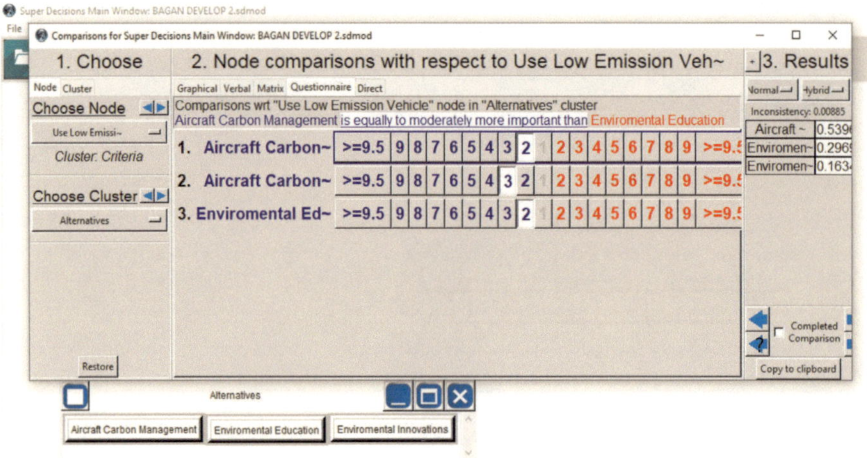
Comparisons for Super Decisions Main Window: BAGAN DEVELOP 2.sdmod
1. Choose
2. Node comparisons with respect to Use Low Emission Veh~
3. Results
Comparisons wrt "Use Low Emission Vehicle" node in "Alternatives" cluster
Aircraft Carbon Management is equally to moderately more important than Enviromental Education
1. Aircraft Carbon~
2. Aircraft Carbon~
3. Enviromental Ed~
Inconsistency: 0.00885
Choose Node
Choose Cluster
Cluster: Criteria
Restore
Copy to clipboard
Aircraft Carbon Management
Enviromental Education
Enviromental Innovations

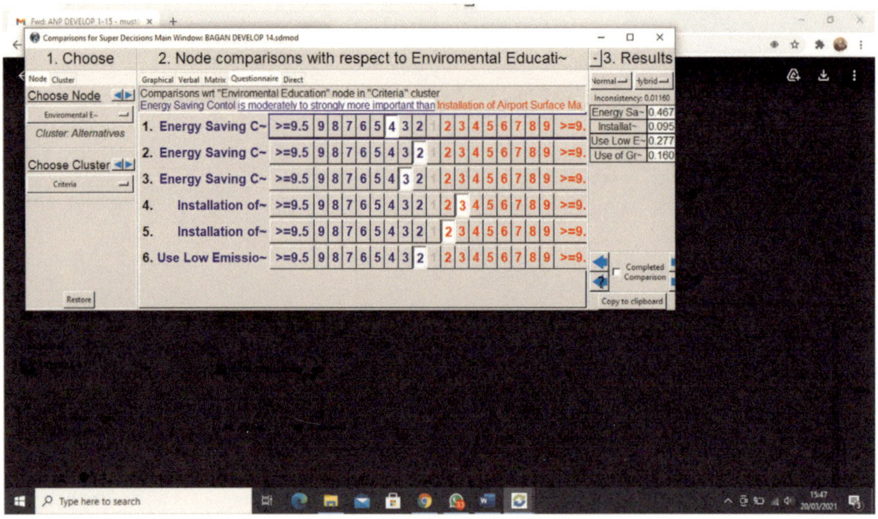
Comparisons for Super Decisions Main Window: BAGAN DEVELOP 14.sdmod
1. Choose
2. Node comparisons with respect to Enviromental Educati~
3. Results
Comparisons wrt "Enviromental Education" node in "Criteria" cluster
Energy Saving Contol is moderately to strongly more important than Installation of Airport Surface Ma
1. Energy Saving C~
2. Energy Saving C~
3. Energy Saving C~
4. Installation of~
5. Installation of~
6. Use Low Emissio~
Inconsistency: 0.01160
Cluster: Alternatives
Type here to search

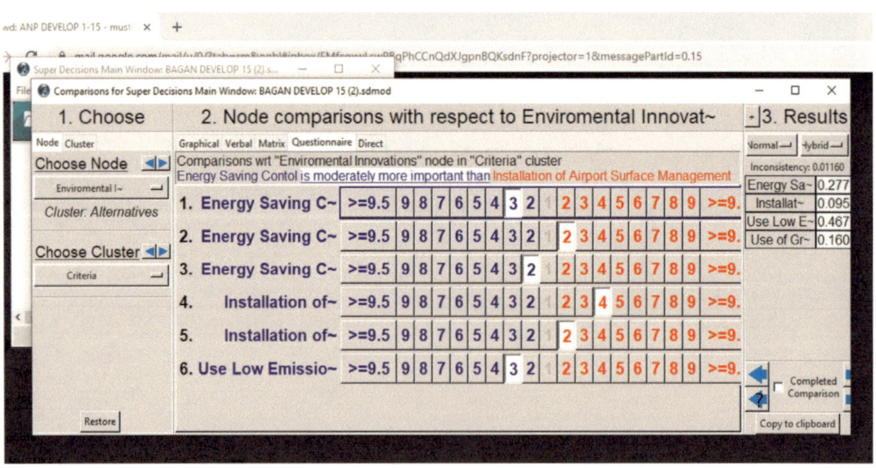
Comparisons for Super Decisions Main Window: BAGAN DEVELOP 15 (2).sdmod
1. Choose
2. Node comparisons with respect to Enviromental Innovat~
3. Results
Comparisons wrt "Enviromental Innovations" node in "Criteria" cluster
Energy Saving Contol is moderately more important than Installation of Airport Surface Management
1. Energy Saving C~
2. Energy Saving C~
3. Energy Saving C~
4. Installation of~
5. Installation of~
6. Use Low Emissio~
Inconsistency: 0.01160
Cluster: Alternatives

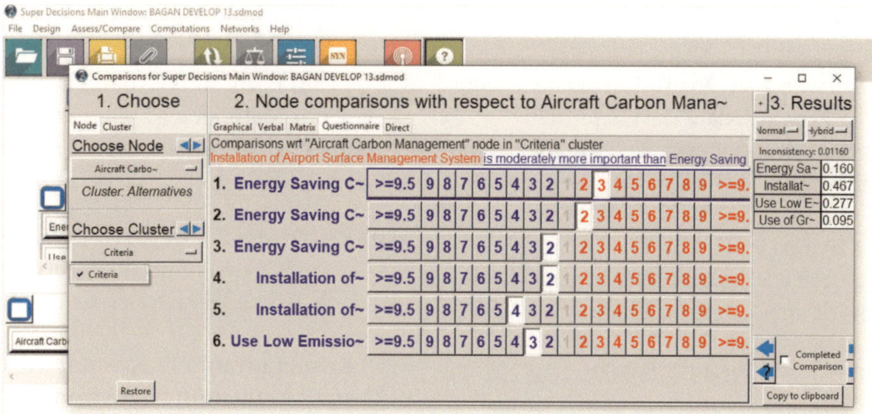

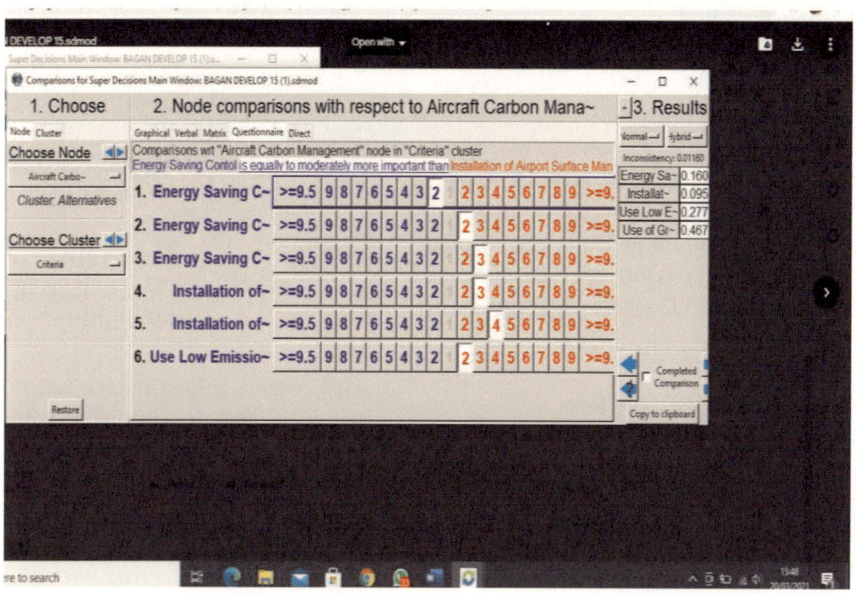

Author details

Mustika Sari* and Reni Dian Octaviani
Institute Transportation and Logistics Trisakti, Jakarta, Indonesia

*Address all correspondence to: mustika0017@gmail.com

References

[1] Purwanta W. Profil Emisi Gas Buang dari Pesawat Udara dari Sejumlah Bandara di Indonesia. Tehnologi Lingkungan. 2015;**16**(1):21-26

[2] Marintseva K, Yun G, Kachur S. Resource allocation improvement in the tasks of airport ground handling operations. Aviation. 2015;**19**(1):7-13. DOI: 10.3846/16487788.2015.1015291

[3] Susanti. The study of human factor of ground handling human resources in Yogyakarta Adi Sucipto airport. Jurnal Perhubungan Udara. 2016;**42**(1):29-42

[4] Hayat B, Asri E, Agnaou A, Fakhrulddin A, Al-humairi A. Ground handling operations management: An agent-based modelling approach. Global journal of science frontier research. Mathematics and Decision Science. 2018; **18**(6):15-22

[5] Airport Cooporative Research Program. Airport Ground Support Equipment (GSE): Emission Reduction Strategies, Inventory, and Tutorial, Appendix B. The National Academies Press; 2012. DOI: 10.17226/22681

[6] Doddy. Reduce emissions. Go Green Aviation. 2012;**2**:2

[7] Indonesia Team. Airport Emissions Reduction Actions in Indonesia through the Establishment of Eco Airport Council and Emissions Inventory. Aviation and Environment (DGCA 55/IP/7/10DGCA); 2018

[8] International Air Transport Association. Airport Handling Manual. 34th ed2014 IATA

[9] Suharto & Eko. Ground Handling. 2nd ed. Raja Grafindo Persada; 2010

[10] Fleuti E. Aircraft Ground Handling Emissions Methodology and Emission Factors Zurich Airport. Zurich Airport; 2014

[11] Kristensen K, Brogaard L, Jacobsen J, Ellerman T. Air Pollution in Airports. 1st ed. Vol. 1. The Danish Eco Council; 2012

[12] Kusnoputranto & Kristanto. Capacity evaluation of environmental noise for airport sustainability through. Research Journal of Science & IT Management. 2014;**3**(3):20-27

[13] Chapman L. Transport and climate change: A review. Journal of Transport Geography. 2007;**5**(5):354-367. DOI: 10.1016/j.jtrangeo.2006.11.008

[14] Rusdiyana & Devi. (2013). Analytic Network Process (Aslam Rusydi (ed.); First, Jan). Smart Consulting

[15] Creswell J. Qualitative Inquiry & Research Design. Choosing Among Five Approaches (Second Edi): Sage Publication; 2007

[16] Saaty T, Vargas L. In: Frederick HS, editor. Decision Making with the Analytic Network Process. Economic, Political, Social and Technological Applications with Benefits, Opportunities, Costs and Risks. New York: Springer; 2006

[17] Percepatan Program Kendaraan Bermotor Listrik Berbasis Baterai (Battery Electric Vehice) Untuk Transportasi Jalan. Vol. 12019

[18] Tax Incentives for Transfer of Ownership of Motor Vehicles on Electric Motorized Vehicles. Vol. 32020

[19] Mariasiu F. Energy Sources Management and Future Automotive

Technologies : International Journal of Energy Economics and Policy. 2012;**2**(4): 342-347

[20] Gapura Angkasa. (2017). *Annual Report Gapura Angkasa Company*. Available from: https://www.gapura.id

[21] Chao L, L. Indicators and evaluation model for analyzing environmental protection performance of airports. Journal of Air Transport Management. 2017;**63**:61-70. DOI: 10.1016/j.jairtraman.2017.05.007

[22] Sari M, Mohamed WMW, Jalil SA. Conceptual model for the implementation of electric ground support equipment In ground handling activities In Indonesia. International Journal of the Analytic Hierarchy Process. 2021;**13**(1):170-182. DOI: 10.13033/ijahp.v13i1.825

Chapter 11

Analytic Hierarchy Process Algorithm Applied to Battery Energy Storage System Selection for Grid Applications

Ana Paula Oening, Débora Cintia Marcilio, Natalie Cristine Scrobot, Danielle de Freitas, Patricio Rodolfo Impinnisi and Cleverson Luiz da Silva Pinto

Abstract

The increasing use of energy storage in batteries has been contributing to the electric sector technological advancement, specially to guarantee of continuous supply in periods of intermittence or low production. In this work the problem of selecting batteries for application in the quality of energy supply is addressed. A methodology to size and classify the best battery technology based on the multicriteria decision-making method AHP (Analytic Hierarchy Process) was developed. The opinions of specialists in the area were considered for the avaliation to four types of criteria: environmental, technological, regulatory and financial. Finally, to evaluate the developed approach a problem of classifying four different battery alternatives (Ventilated Stationary Tubular Lead Acid, Lead-Carbon Acid, Ventilated Stationary Lead Acid, and Lithium Ion LFP) was tested and the result pointed to the Ventilated Stationary Lead Acid technology as the best alternative. In addition, analyses using theoretical scenarios with preferences in certain criteria were performed. The methodology proved applicable, considering that with the input data provided by the decision maker, it is possible suggest the best technological alternative for a given purpose. The simulations carried out suggest that the methodology has potential to evolve and act in a real situation.

Keywords: energy storage system, battery, multicriteria decision-making algorithm, analytic hierarchy process, Brazilian electrical system

1. Introduction

The supply of energy, in times of intermittency or low production, can be assured through energy storage systems (ESS). Efficient management of energy storage is crucial in the process of achieving a balance between power quality, efficiency, costs, and environmental constraints [1]. In Brazil, for example, it is estimated that by 2023 its market will use around 95 GWh in energy storage systems, and this number corresponds to 50% of all installed capacity in the world at the end of 2015 [2].

ESS, in particular battery energy storage systems, are the most frequently installed options to facilitate the use of renewable energy [3], and batteries have been maturing for several decades and increasing their reliability and durability. In Germany, the market for stationary battery storage systems is growing rapidly compared to pumped storage systems [4]. Taking into account the expected increase in consumption of this technology for the coming years, it is expected, as a reflection of its large-scale use, a gradual reduction in its acquisition and implementation costs.

The choice of batteries for energy storage is a complex decision, as it needs to take into account several technical, economic, regulatory, and environmental criteria. In this sense, the multicriteria decision-making (MCDM) methods can serve as a decision support tool.

Conventional MCDM methods are frequently used in the energy field [5] in a wide range of problems [6–12]. In [11], the combination of analytic hierarchy process (AHP) and Vise Kriterijumska Optimizacija I Kompromisno Resenje (VIKOR) methods are used to select the best solution for electrical power supply in remote rural locations. The evaluation and classification of renewable energy sources are addressed as an MCDM problem in [10], based on the theory of gray systems. The problem of selecting investment projects for solar thermal power plants addressed in [12] uses the AHP and Analytic Network Process (ANP) methods.

More specifically in the energy storage problem, we also have the applicability of MCDM methods [1, 13–16]. In [16], an attempt is made to determine the most appropriate among six energy storage alternatives, considering four main criteria and 16 subcriteria. For this, a hybrid methodology is used that combines AHP and Technique for Order Preference by Similarity to Ideal Solution (TOPSIS) based on type 2 fuzzy sets. In [15], a decision support tool is proposed for the selection of energy storage alternatives. Considering a multi-objective optimization approach based on the augmented ϵ constraint method meets the technical constraints, economic, and environmental objectives of the problem. The AHP method combined with fuzzy logic is used in [1] for the selection of energy storage considering criteria of efficiency, load management, technical maturity, costs, environmental impact, and energy quality. In [5], five different battery alternatives (lead-acid, lithium-ion, vanadium redox flow battery, sodium-nickel chloride, and sodium-sulfur) are evaluated using Pythagorean Fuzzy sets combined with AHP-TOPSIS.

A literature review of the main MCDM approaches in evaluating ESS is provided in [3]. As seen earlier, among the multicriteria decision-making methods, we have the AHP method created by the mathematician Thomas L. Saaty [17, 18]. This approach consists of evaluating criteria through a measurement scale, organizing them in a hierarchical structure [18]. Since this is a flexible decision support method, which can take into account objective and subjective criteria by comparing pairs of criteria [19], in the subjective criteria we can consider the opinion of specialists in an area of knowledge. Its use in its original form or in hybrid approaches has applications in several problems in the energy sector [8, 20–26].

Given the growing demand for ESS technologies, especially batteries, here we develop a multicriteria analysis methodology based on the AHP method, which is one of the most traditional MCDM methods used in a wide range of energy field problems. The main objective of this approach through AHP is to help choose the best battery technology for energy storage, considering for this, technical, regulatory, environmental, and economic criteria and subcriteria through the opinions of specialists in the area.

This chapter is organized as follows. In Section 2, we make a brief summary of the AHP method. In Section 3, we present the methodology developed to choose the best battery technology for energy storage using the AHP method. To validate the developed methodology, we performed numerical experiments in Section 4, considering an analysis of a scenario based on expert opinion and another analysis based on theoretical scenarios. Finally, in Section 5, we present our conclusions.

2. Brief review of the AHP method

Decision theory is the set of techniques that help decision-makers to recognize the particularities of a problem and structure it. In this context, MCDM refers to determining the best option in relation to multiple and often conflicting criteria [27].

Among the traditional methods of MCDM, we have the AHP method which has already been applied in various problems in the energy field, such as renewable energy [26, 28–30], energy storage [3, 13, 21, 31–33], site selection for wind farms [22, 34–36], among others [37–40].

This method was created by the mathematician Thomas L. Saaty in the 70s [17, 18], being a method used to solve problems with multiple criteria and which emphasizes modeling according to the knowledge of specialists in the area. This method considers quantitative and qualitative aspects and can be used both to identify the best possible alternative and also to list the alternatives according to a ranking of priorities.

One of the advantages of using the method is the consideration of the experience of the decision-maker who can intuitively assign weights to the relevant criteria by comparing them pair by pair. **Figure 1** describes the steps of this method.

First, the alternative solutions to the problem are listed, and the criteria and subcriteria for choosing the alternatives are listed. Then, based on the judgment of the decision-makers, scores are assigned to each criterion following the Saaty scale [17, 18], as described in **Table 1**.

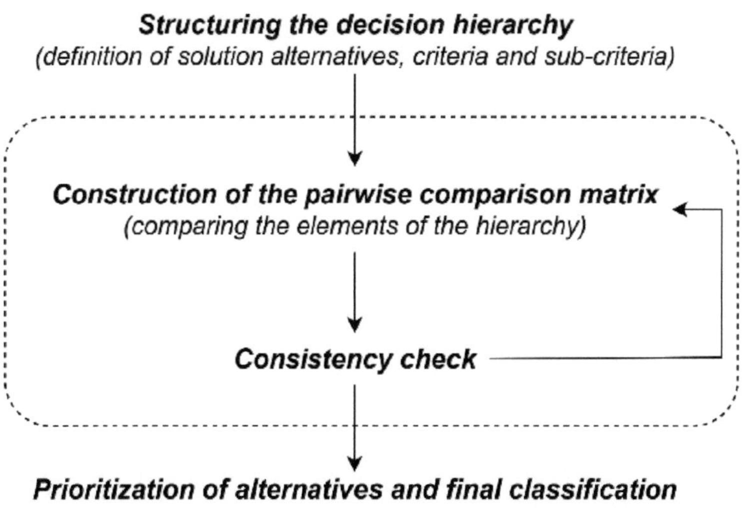

Figure 1.
Steps of the method AHP.

Intensity of importance	Description
1	Equal importance
3	Moderate importance of one over another
5	Essential or strong importance
7	Very strong importance
9	Extreme importance
2, 4, 6, 8	Intermediate values between the two adjacent judgments
Reciprocals	If judgment of factor i compared with factor j is a_{ij}, then judgment of factor j compared with factor i is $a_{ji} = 1 = a_{ij}$.

Table 1.
The Saaty rating scale for deciding relative importance of factor i compared to factor j [17].

Each criterion/subcriteria are judged pair by pair, and this judgment is represented through a square decision matrix $A = \left(a_{ij}\right) \in \mathbb{R}^{n\times n}$, where n is the number of evaluations of criteria/subcriteria. To calculate the consistency of the judgments made, we first normalize the decision matrix, according to Eq. (1):

$$A_{norm} = \overline{a}_{ij} \in \mathbb{R}^{n\times n}, \text{such that } \overline{a}_{ij} = {}^{a_{ij}}/{\sum_{i=1}^{n} a_{ij}} \tag{1}$$

must satisfy the equation Eq. (2):

$$Av = \lambda v \tag{2}$$

where A represents the squared decision matrix, λ is the eigenvalue with the largest value of A, and v is its corresponding eigenvector.

Then, the consistency ratio (CR) of each decision-maker's preferences is calculated, according to the Eq. (3):

$$CR = \frac{(\lambda_{max} - n)}{(n-1)} \bullet \frac{1}{RCI} \tag{3}$$

where RCI is the random consistency index, obtained from **Table 2**, n is the number of evaluated criteria, and λ_{max} is the maximum eigenvalue of A.

If the value of CR is less than 0.1 we say that matrix A is consistent, otherwise the matrix is inconsistent and decision-makers must repeat their evaluations until a satisfactory CR is obtained.

After this brief explanation about the AHP theory, in the next section, we will present the methodology developed for sizing batteries and show the approach developed for the application of the AHP method, which uses some of the information obtained through sizing.

n	1	2	3	4	5	6	7	8	9	10
RCI	0.00	0.00	0.52	0.89	1.11	1.25	1.35	1.40	1.45	1.49

Table 2.
Average random consistency index [41].

3. Approach developed for the application of the AHP method

In order to fulfill the objective of listing the battery alternatives, the "Input data" stage concerns the collection of essential data for the other calculations, originating from the region where the battery bank will be allocated, the demand, the conditions of use of the application, and the information provided by the manufacturers. In the dimensioning stage, the application data are combined with the data provided by the manufacturers, and the results of this stage alone provide a range of information on the different batteries evaluated, some of which will serve as input for the AHP method. In the "Application of the AHP method" stage, the batteries are classified according to each of the considered criteria and subcriteria. As a result, AHP lists battery alternatives in a priority ranking (**Figure 2**).

3.1 Input data of the problem

One of the objectives of this work is the classification of different battery technologies to form a bank that meets a certain demand. Since this ordering is established considering the preference of certain qualification criteria. Therefore, input data are necessary for the development of the methodology, namely:

a. average temperature of the region: provided in C°, being necessary for calculating the time of use of the batteries;

b. energy demand: supplied in kWh, used to calculate the energy of the battery bank in the sizing.

c. operating regime: in this regard, we provide the period considered (hours, days, etc.), number of cycles per period, and the duration of each cycle per hour, and this information is used to determine the time of use of the battery bank.

d. battery database: these are made available by the manufacturers and used in the sizing calculation.

3.2 Battery sizing

In the methodology developed for dimensioning the batteries, the application data entered by the user and the information contained in the database are combined. The variables that make up the sizing are gravimetric density, volumetric density, optimal Depth of Discharge (DOD), battery bank energy, number of containers, and battery usage time. Determining the sizing variables provides the user with a range of information on the different battery alternatives evaluated, and some of this information serves as input for the AHP method.

Therefore, the design calculation was divided into six steps, as shown in the following algorithm. In Step 1, we described the calculation of gravimetric energy density. In Step 2, we present the calculation of the volumetric energy density. In Step 3, we show the methodology adopted to determine the optimal DOD. The calculation for energy from the battery bank is performed in Step 4. In Step 5, we establish the logic for defining the number of containers for storing the battery bank, taking two options for containers with volumes of 33 m^3 and 67 m^3, named type I and type II,

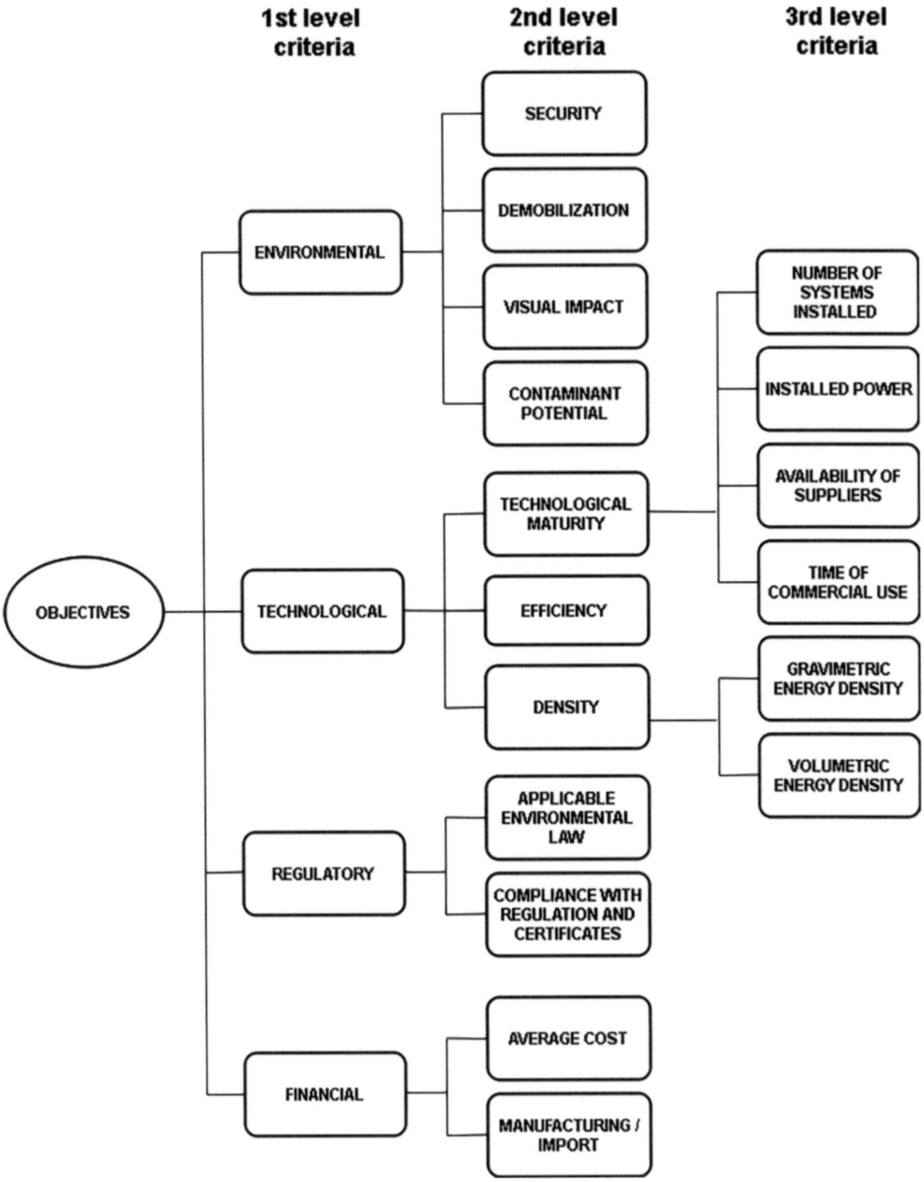

Figure 2.
Overview of the developed methodology.

respectively. Since, we consider only 60% of the total capacity of the container, since the remaining space is destined for the control and refrigeration systems. Finally, the calculation of the battery usage time is presented in Step 6.

Battery sizing algorithm

Step 1 Gravimetric Energy Density

Require: Capacity (Ah), Voltage (V) and Weight (kg)

For each type of battery technology **do**

$$\text{Gravimetric density} \leftarrow \frac{\text{Capacity} \times \text{Voltage}}{\text{Monoblock Weight}}$$

End For

Step 2 Volumetric Energy Density

Require: Capacity (Ah), Voltage (V) and Volume (m^3)
For each type of battery technology **do**
$\text{Volumetric density} \leftarrow \frac{\text{Capacity} \times \text{Voltage}}{\text{Monoblock Volume}}.$
End For

Step 3 DOD great

Require: Energy Throughput (kWh) and DOD (percentage)
For each type of battery technology **do**
Calculate the $\max_{\text{DOD}} (\text{Energy Throughput}).$
If you have a maximum **then**
great DOD $\leftarrow$ DOD*, where DOD* is the solution obtained
Else
great DOD $\leftarrow$ manufacturer's DOD.
End If
End For

Step 4 Battery bank power

Require: Optimal DOD (percentage) and Demand (kWh)
For each type of battery technology **do**
$\text{Bank Energy} \leftarrow \frac{\text{Demand}}{\text{Optimal DOD}}.$
End For

Step 5 Number of containers

Require: Bank Volume (m^3) and Effective Container Capacity (m^3)
For each type of battery technology **do**
$\text{Type I Containers} \leftarrow \left(\frac{\text{Bank Volume}}{0.6 \times \text{Type I Container Capacity}}\right).$
$\text{Type II Containers} \leftarrow \left(\frac{\text{Bank Volume}}{0.6 \times \text{Type II Container Capacity}}\right).$
Optimal number of containers $\leftarrow$ min(Type I containers, Type II containers)
End For

Step 6 Battery usage time

Require: Number of optimal DOD cycles and number of cycles per period
For each type of battery technology **do**
$\text{Usage time} \leftarrow \frac{\text{Number of optimal DOD cycles}}{365 \times \text{Number of cycles per period}}.$
End For

3.3 Application of AHP

The methodology developed was divided into four phases: (i) definition of alternatives, (ii) definition of criteria and subcriteria, (iii) pairwise comparison and consistency check, and (iv) final classification. The methodology for phases (i) to (iii) is presented in this section. Phase (iv) consists of the calculations for prioritizing the alternatives and final classification, and these results are presented in Chapter 4.

3.3.1 Definition of battery alternatives

For the developed approach, four types of battery technologies were chosen, according to **Table 3**.

Batteries 1, 2, and 3 are lead-acid, one of the most used and developed technologies in the world. It has a low acquisition, installation, and maintenance cost and competes with lithium-ion batteries when weight and volume are not decisive factors. These batteries have the following characteristics: fast response time, low percentage of self-discharge, fully recyclable, presenting risks of corrosion, exposure to toxic agents, leakage, and potential for contamination. However, there are still limitations such as energy density and malfunction at low temperatures [42].

Battery 4 is iron phosphate lithium ion, and this technology features low weight, good storage capacity, absence of memory effect, high energy density, risk of explosion, exposure to toxic agents, contamination by lithium, phosphorus, and organic solvents [43].

3.3.2 Definition of criteria and subcriteria

Based on previous studies and the experience of the researchers who developed this work, four criteria were defined for choosing battery alternatives, namely: environmental, technological, regulatory and financial, and subcriteria were also defined according to **Figure 3**. The main idea for these criteria is to consider different aspects that can influence this type of decision-making. Accordingly, a breakdown of the criteria and subcriteria is provided below.

3.3.2.1 EC environmental criteria

EC1. Security: It refers to sporadic environmental risks, whose indicators are associated with potential damage. The risks are classified based on the sum of scores distributed as follows: 1 for the risk of corrosion, 2 for the risk of exposure to toxic agents, and 3 for the risk of explosion and fluid leakage. In this subcriterion, the lower the score obtained, the greater the preference for drums.

EC2. Demobilization: It refers to the recycling of batteries according to the material they are made of and the types of waste generated at the end of the equipment's useful life.

N°	Battery alternatives	Manufacturer	Model
1	Ventilated Stationary Tubular Lead Acid (OPzS)	Fulguris	3OPzs150
2	Lead-Carbon Acid (PbC)	Narada	12RECX120
3	Ventilated Stationary Lead Acid	Moura Clean	12MF175
4	Lithium-Ion LFP (Iron Phosphate)	UniPower	UP-LFP 4875

Table 3.
Battery technology alternatives.

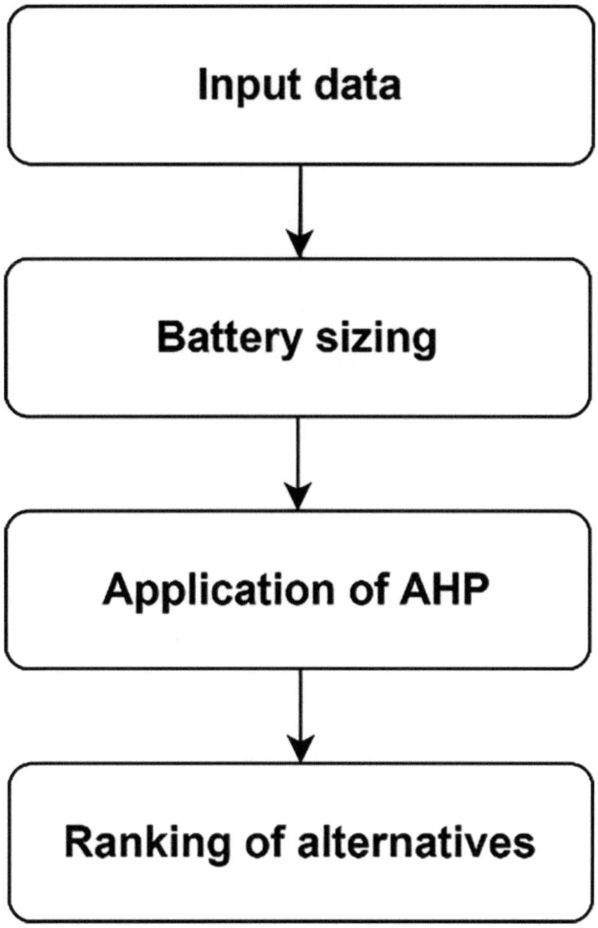

Figure 3.
Hierarchical structure of the problem criteria.

EC3. Visual Impact: It refers to the visual impact caused by the size of the containers that will be used to support the battery bank, considering that the larger the container, the lesser the importance of the technology.

EC4. Contaminant Potential: It refers to the environmental damage on the biota caused by the chemical components of the batteries. The following damage classes were defined: low when the contaminating elements reduce the quality of fauna and flora and even affect local ecosystems, medium when they can reduce reproduction or biological survival, and high or very high when they cause damage that can lead to death of organisms and the impossibility of reproduction.

3.3.2.2 TC. Technological criteria

TC1. Technological Maturity

TC11. Number of Systems Installed: It refers to the number of systems installed worldwide for each battery technology this information is taken from [44]. In this third-level criterion, the greater the number of projects, the greater the importance of technology.

TC12. Installed Power: It refers to the total amount of installed power in projects deployed in the world for each technology, and the information was taken from [44]. In this subcriterion, the greater the installed capacity, the greater the importance of technology.

TC13. Availability of suppliers: With regard to the number of suppliers available for each type of technology, this information was obtained from [44]. In this subcriterion, the greater the number of suppliers, the greater the importance of technology.

TC14. Time of Commercial Use: The time in which the technology is commercially available is evaluated, that is, the longer the time, the greater the degree of importance of the technology.

TC2. Efficiency: Maximum percentage charge and discharge of stored energy with minimal decrease in battery life. When evaluating this criterion, it is defined that the higher the percentage, the greater the degree of importance.

3.3.2.3 TC3. Density

TC31. Gravimetric Energy Density: Amount of energy stored per unit mass, expressed in Wh/kg. For density criteria, the higher the density, the greater the degree of importance. The value is numeric and undergoes a normalization process to generate the score for this subcriterion.

TC32. Volumetric energy density: Amount of electrical charge stored per volume; in this case, the greater the density, the greater the degree of importance, being measured in Wh/mm^3 As this criterion is numerical, the values undergo normalization to acquire the score for each technology.

3.3.2.4 RC. Regulatory criteria

RC1. Applicable Environmental Law: It refers to the existence of legal provisions for the analysis of technologies with regard to the operation, since the existence of legislation gives legal certainty to contracts. The adaptability of the legislation can be considered by the absence of specific legal provisions, and the degree of importance is increased as the alternative has valid and applicable legislation.

RC2. Compliance with regulations and certificates: It is evaluated by the existence or not of norms and/or specific laws for a certain type of technology. The degree of importance increases due to the existence of specific norms and laws.

3.3.2.5 FC. Financial criteria

FC1. Average Cost: It refers to the average cost of purchase, installation, and maintenance of the battery system; in this criterion, the higher the average cost, the lesser its importance. If there are exact values obtained through budgets, this value can be used and goes through a normalization process to obtain the score for each technology.

FC2. Manufacturing/Import: It concerns the existence of domestic or foreign battery factories. The degree of importance is increased if the company has a national domicile, as it excludes possible difficulties related to importation.

With the exception of the security and density subcriteria, we can associate ranges of importance to each subcriteria according to **Table 4**. Note that if the average cost

Subcriteria	Feature evaluated	Small	Great	Very large	Absolute
EC2	Recycling	Disposable	Partially recycling	Fully recycling	—
ec3	Number of containers	More than 2 type II containers	2 type II containers	1 type II containers	2 type I containers
EC4	Damage to the biota	Very serious	Serious	Average	Little
TC11	Projects in the world	[0,10[	[10,40[	[40,80[	≥80
TC12	Energy (MW)	[0,10[	[10,50[	[50,100[	≥100
TC13	Number of suppliers in the world	[0,5[	[5,20[	[20,40[	≥40
TC14	Usage time worldwide (years)	[0,10[	[10,30[	[30,60[	≥60
TC2	Efficiency percentage	[0,0.8[	[0.8,0.9[	[0.9,0.95[	≥0.95
RC1	Applicability of law	Not applicable	Law can be adapted	Established law but with difficult application	Established and applicable law
RC2	Compliance with regulations and issuing of certificates	There is no norm and law	Is there a rule	There is rule and there is law	—
FC1	Average cost ($/MWh)	> 1000	] 500,1000]	]100,500]	≤100
FC2	Domestic or foreign manufacturing	Foreign	Domestic	Foreign	Domestic

Table 4.
Level of importance per characteristic evaluated in each subcriterion.

subcriterion does not have a budgeted value, studies can be used to estimate this value, adopting the importance ranges defined in **Table 4**.

3.3.3 Pairwise comparison and consistency check

After defining the alternatives and the hierarchical model of the problem, according to the AHP method, the pairwise comparisons begin to determine the level of importance between the criteria and subcriteria. For that, a questionnaire was developed with comparison matrices between the criteria and subcriteria, separated by type and level. For each comparison matrix, a pairwise evaluation must be performed using the Saaty scale (**Table 1**).

To carry out these evaluations, the questionnaire was sent to 12 professionals with degrees in the areas of Electrical, Environmental, Chemical, Materials, Forestry, in addition to the areas of Biology and Ecology.

From the responses of the questionnaire, the process of normalization and consistency of the scores is assigned by the specialists to validate the interdependence of the criteria and subcriteria (Section 2). The experts' inconsistent assessments were dismissed. In sequence, to define the final weights to be used, we considered the geometric mean of the normalized scores of the eight specialists with consistent responses.

4. Numerical experiments

We performed numerical experiments for the problem of classifying different battery alternatives for energy storage in a region of Brazil, with an application demand of 50 kWh, two cycles of battery use per day, where each cycle lasts 10 hours, and also a temperature of 25C°.

4.1 Dimensioning of the chosen alternatives

Based on the input data of the problem, on the manufacturer's catalog and the sizing methodology developed in Section 3, the characteristics of the batteries were obtained and calculated according to **Table 5**.

4.2 Evaluation of the batteries according to the criteria

Through the results obtained in the dimensioning phase and also through the manufacturers' data, in **Table 6**, we describe the results of each battery in relation to each criterion.

4.3 Application of the AHP method

To rank the different battery technologies, we use the AHP method. For this study, we considered two types of analysis. The first considers the view of specialists in the areas of engineering, environment, and chemistry regarding the analyzed criteria. The second considers theoretical cases where scenarios are generated considering the preference in each of the environmental, technological, regulatory, or financial criteria, and even a neutral scenario where all criteria have the same level of importance.

4.3.1 Expert scenario analysis

As presented in section 2, based on the consistent evaluations of the eight experts, the weights for each criterion and subcriteria were defined, according to **Table 7**. This set of weights was named Expert scenario.

Considering the weights defined by the expert's answers (**Table 7**) and the weights defined through the ranges of interest (**Table 4**) together with the information of the batteries made available (**Table 6**), the final score of each heat in each criterion is provided in **Table 8**, and also the final priority of the heats.

Battery alternatives	Gravimetric energy density (Wh/kg)	Volumetric energy density (Wh/m³)	Optimal DOD (%)	Battery pack energy supplied (kWh)	Containers number	Battery usage time (years)
1	18.75	3.82E+04	54.43	91.9	1 of type I	3.53
2	23.53	6.27E+04	25	200	1 of type I	13.13
3	35.42	5.55E+04	20	250	1 of type I	2.47
4	87.80	1.15E+05	64	78.12	1 of type I	6.19

Table 5
Result of sizing battery alternatives.

Criteria	Battery 1	Battery 2	Battery 3	Battery 4
EC1	6	6	6	5
EC2	Recyclable	Recyclable	Recyclable	Disposable
EC3	1 container of type I	1 container of type I	1 container of type I	1 container of type I
EC4	Little damage	Serious damage	Average damage	Little damage
TC11	45	3	69	112
TC12 (MW)	61.4	1.0	61.4	174.6
TC13	[40,80[	[5,20[	[40,80[	[40,80[
TC14	≥ 60	[10,30[	≥ 60	[30,60[
TC2	0.85	0.92	0.85	0.98
TC31 (kWh/kg)	0.01875	0.02353	0.03542	0.08780
TC32 (kWh/m^3)	3.82E+04	6.27E+04	5.55E+04	1.15E+05
RC1	Law enacted and applicable to technology	Law can be adapted	Law enacted and applicable to technology	Law can be adapted
RC2	Standard only	No standard	Standard only	Standard only
FC1 ($/MWh)	]500,1000]	]500,1000]	]500,1000]	> 1000
FC2	Brazil	Abroad	Brazil	Abroad

Table 6.
Characteristics of each battery regarding the criteria.

From the perspective of the environmental criterion, the indicated technology is that of lithium ions; however, the difference between the results of the four technologies was not substantial. This alternative is also the most adequate regarding the technological criterion, in this case with a wide advantage in relation to the others. As for the regulatory and financial criteria, lead-acid (OpzS) and vented lead-acid batteries obtained the same results, since the two technologies have exactly the same input data for the associated subcriteria as show in **Table 6**.

Finally, with the assumptions adopted in this study and the four technologies evaluated, the final results are presented in **Table 9**. It appears that the overall result pointed to the Ventilated Stationary Lead-Acid technology as the best alternative for this study, that is, 27.92% priority over the others.

The results of the final priorities of technologies 1, 3, and 4 are very similar (around 2% off difference), so they can be considered technically equivalents and small changes in weights may result in different classifications.

4.3.2 Analysis of theoretical scenarios

With the AHP weights established, we can create other scenarios, which can explain different user preferences. For example, assuming a greater preference for a

1st level criteria	(%)	2nd level criteria	(%)	3nd level criteria	(%)
EC	24.64	EC1	37.39		
		EC2	20.20		
		EC3	5.35		
		EC4	37.09		
TC	29.64	TC1	19.04	**TC11**	12.92
				TC12	33.28
				TC13	37.07
				TC14	16.72
		TC2	36.90		
		TC3	44.04	**TC31**	54.65
				TC32	45.35
RC	21.07	RC1	51.67		
		RC2	48.33		
FC	24.64	FC1	61.45		
		FC2	38.55		

Table 7
Weights for each criterion and subcriteria based on consistent expert responses.

Battery	Environmental	Technological	Regulatory	Financial
1	0.2371	0.1700	**0.3415**	**0.3267**
2	0.2371	0.1819	0.1186	0.2433
3	0.2527	0.2176	**0.3415**	**0.3267**
4	**0.2730**	**0.4305**	0.1985	0.1033

Table 8
Notes for each battery against the criteria.

certain criterion and a lower weight for the others, or leaving them all with the same decision-making power, in this way the decision-maker indicates his priority.

To create these scenarios, we changed the weights given to the first-level criteria (Environmental, Technological, Regulatory, and Financial), while the rest of the weights are defined according to the expert's scenario (**Table 7**).

In this way, we generate five different scenarios, namely: environmental, technological, regulatory, financial, and neutral. The weights for the criteria of each scenario can be seen below in **Table 10**, with the criterion with the highest weight highlighted in bold.

Finally, considering the assumptions adopted in this study and the four technologies evaluated, **Table 11** shows the final result of the AHP for each scenario. The general result presents the priority ranking of the batteries, with the Stationary Vented Lead-Acid technology being the best alternative for this application, among the four of the five analyzed scenarios. The evaluation of these scenarios is interesting to verify the performance of technologies considering different strategies.

	Battery	Final priority
1	Ventilated Stationary Tubular Lead Acid (OPzS)	26.13%
2	Lead-Carbon Acid (PbC)	19.73%
3	Ventilated Stationary Lead Acid	**27.92%**
4	Lithium Ion LFP (Iron Phosphate)	26.22%

Table 9
Final result of the AHP for each battery technology.

		Scenario				
		Environmental	**Technological**	**Regulatory**	**Financial**	**Neutral**
Criteria	EC	**0.4000**	0.2000	0.2000	0.2000	**0.2500**
	TC	0.2000	**0.4000**	0.2000	0.2000	**0.2500**
	RC	0.2000	0.2000	**0.4000**	0.2000	**0.2500**
	FC	0.2000	0.2000	0.2000	**0.4000**	**0.2500**

Table 10
Weights for the 1st level criteria in each scenario

Battery	Scenarios priorities				
	Environmental	**Technological**	**Regulatory**	**Financial**	**Neutral**
1	26.25%	24.90%	28.33%	28.04%	26.88%
2	20.36%	19.26%	17.99%	20.48%	19.52%
3	**27.82%**	27.12%	**29.60%**	**29.30%**	**28.46%**
4	25.57%	**28.72%**	24.08%	22.18%	25.14%

Table 11
Final AHP result for each battery technology in each scenario.

5. Conclusions

The Brazilian Power Sector is preparing the introduction of battery energy storage in its distribution lines for energy quality control. The success or failure of this new technology (from a financial and technical standpoint) depends on many factors. Different kinds of battery technologies have advantages and disadvantages depending on the operational systems regimes adopted. Different geographical locations, accessibility, spare parts availability, and other related factors may play a significant role in giving priority to one technology rather than another, and even environmental restrictions and the local regulatory framework can affect the results. The AHP algorithm is designed to take into account all these factors. It should be pointed out that the results depend a great deal on the evaluations of the specialists. In view of this, as many experts as possible must be consulted for each evaluated criteria and subcriteria to overcome this subjectivity.

In this work, we present a methodology to size and classify different battery alternatives, based on the AHP method. For this, considering the opinions of experts in the field, grades were assigned to different criteria and subcriteria, defined from the main characteristics of batteries in the segments: environmental, technological, regulatory, and financial.

Finally, to validate the developed methodology, we approach a problem of classifying four different types of battery technologies (Ventilated Stationary Tubular Lead Acid, Lead-Carbon Acid, Ventilated Stationary Lead Acid, and Lithium Ion LFP). In the study, two types of analysis were performed. The first analysis considered the specialists' view about the analyzed criteria, and the study pointed out that the best alternative for the application was the Ventilated Stationary Lead-Acid battery.

The second analysis considered theoretical cases, with scenarios generated assuming a preference for each of the first-level criteria, and even a neutral scenario where all criteria have the same level of importance. In the results, the Ventilated Stationary Lead-Acid battery was the best alternative among the four of the five analyzed scenarios.

It is important to note that the useful life of the alternatives was used to calculate the batteries dimensioning; however, it was not considered as a criterion in the AHP method. In the future work, it is expected to use the useful life with the real prices of the technologies to replace the average price subcriteria.

The results of the method must be evaluated by specialists to validate the choice, since they have greater knowledge of the particularities of each technology. The advantage of using AHP for this problem is the consideration of subjective criteria such as environmental and regulatory criteria, which cannot be used directly in the calculation of the cost-benefit of technologies.

Acknowledgements

This research is been funding by R&D project PD-2866-0444/2016 proposed by Parana State Energy Company—Copel, under the auspices of the R&D Program of the Brazilian Electricity Regulatory Agency—ANEEL. The authors are also indebted to National Council for Scientific and Technological Development—CNPq for the subsidies for importation through law 8010/1990, I.I. 17/3098290-0.

Author details

Ana Paula Oening[1*], Débora Cintia Marcilio[1], Natalie Cristine Scrobot[1], Danielle de Freitas[1], Patricio Rodolfo Impinnisi[1] and Cleverson Luiz da Silva Pinto[2]

1 LACTEC, Curitiba, Brazil

2 COPEL, Curitiba, Brazil

*Address all correspondence to: ana.oening@lactec.com.br

References

[1] Barin A, Canha LN, Abaide AR, Magnago KF, Wottrich B, Machado RQ. Multiple criteria analysis for energy storage selection. Energy and Power Engineering. 2011;**3**(04):557

[2] Bueno A, Brandão C. Visão geral de tecnologia e mercado para os sistemas de armazenamento de energia elétrica no brasil. Associação Brasileira de Armazenamento e Qualidade de Energia. 2016;**1**:1-62

[3] Baumann M, Peters J, Weil M. Exploratory multicriteria decision analysis of utility-scale battery storage technologies for multiple grid services based on life-cycle approaches. Energy Technology. 2020;**8**:1-19

[4] Figgener J, Stenzel P, Kairies K-P k, Linßen J, Haberschusz D, Wessels O, et al. The development of stationary battery storage systems in Germany – A market review. Journal of Energy Storage. 2020;**29**:101153. DOI: 10.1016/j.est.2019.101153

[5] Bulut M, Özcan E. A novel approach towards evaluation of joint technology performances of battery energy storage system in a fuzzy environment. Journal of Energy Storage. 2021;**36**:102361. DOI: 10.1016/j.est.2021.102361

[6] Wu Y, Zhang T, Gao R, Wu C. Portfolio planning of renewable energy with energy storage technologies for different applications from electricity grid. Applied Energy. 2021;**287**:116562. DOI: 10.1016/j.apenergy.2021.116562

[7] Ahmadi SHR, Noorollahi Y, Ghanbari S, Ebrahimi M, Hosseini H, Foroozani A, Hajinezhad A. Hybrid fuzzy decision-making approach for wind-powered pumped storage power plant site selection: A case study. Sustainable Energy Technologies and Assessments. 2020;**42**: 100838. DOI: 10.1016/j.seta.2020.100838.385

[8] Bohra SS, Anvari-Moghaddam A, Mohammadi-Ivatloo B. Ahp-assisted multi-criteria decision-making model for planning of microgrids. In: IECON 2019 - 45th Annual Conference of the IEEE Industrial Electronics Society. 2019. pp. 4557-4562. DOI: 10.1109/IECON.2019.8926887

[9] Nzotcha U, Kenfack J, Blanche Manjia M. Integrated multi-criteria decision making methodology for pumped hydro-energy storage plant site selection from a sustainable development perspective with an application. Renewable and Sustainable Energy Reviews. 2019;**112**:930-947. DOI: 10.1016/j.rser.2019.06.035

[10] Çelikbilek Y, Tüysüz F. An integrated grey based multi-criteria decision making approach for the evaluation of renewable energy sources. Energy. 2016;**115**:1246-1258. DOI: 10.1016/j.energy.2016.09.091

[11] Rojas-Zerpa JC, Yusta JM. Application of multicriteria decision methods for electric supply planning in rural and remote areas. Renewable and Sustainable Energy Reviews. 2015; **52**:557-571. DOI: 10.1016/j.rser.2015.07.139

[12] Aragonés-Beltrán P, Chaparro-González F, Pastor-Ferrando J-P, PlaRubio A. An ahp (analytic hierarchy process)/anp (analytic network process) based multi-criteria decision approach for the selection of solar-thermal power plant investment projects. Energy. 2014; **66**:222-238. DOI: 10.1016/j.energy.2013.12.016

[13] Liu Y, Du JL. A multi criteria decision support framework for renewable energy storage technology selection. Journal of Cleaner Production. 2020;**277**:122183. DOI: 10.1016/j.jclepro.2020.122183

[14] Zhao H, Guo S, Zhao H. Comprehensive performance assessment on various battery energy storage systems. Energies. 2020;**11**(10):2841. DOI: 10.3390/en11102841

[15] Li L, Liu P, Li Z, Wang X. A multi-objective optimization approach 415 for selection of energy storage systems. Computers & Chemical Engineering. 2018;**115**:213-225. DOI: 10.1016/j.compchemeng

[16] B. Özkan, Íhsan Kaya, U. Cebeci, H. Ba¸slıgil, A hybrid multicriteria decision making methodology based on type-2 fuzzy sets for selection among energy storage alternatives, International Journal of Computational Intelligence Systems 8 (5) (2015) 914–927. doi: 10.1080/18756891.2015.

[17] Saaty TL. A scaling method for priorities in hierarchical structures. Journal of Mathematical Psychology. 1977;**15**(3):234-281. DOI: 10.1016/0022-2496(77)90033-5

[18] Saaty TL. How to make a decision: The analytic hierarchy process. European Journal of Operational Research. 1990;**48**(1):9-26. DOI: 10.1016/0377-2217(90)90057-I

[19] Kushal TRB, Illindala MS. A decision support framework for resilience-oriented cost-effective distributed generation expansion in power systems. In: 2020 IEEE/IAS 56th Industrial and Commercial Power Systems Technical Conference (I CPS). Las Vegas, USA. 2020. pp. 1-8. DOI: 10.1109/ICPS48389.2020.9176823.440

[20] Baumann M, Weil M, Peters JF, Chibeles-Martins N, Moniz AB. A review of multi-criteria decision making approaches for evaluating energy storage systems for grid applications. Renewable and Sustainable Energy Reviews. 2019;**107**:516-534. DOI: 10.1016/j.rser.2019.02.016

[21] Paul S, Nath AP, Rather ZH. A multi-objective planning framework for coordinated generation from offshore wind farm and battery energy storage system. IEEE Transactions on Sustainable Energy. 2020;**11**(4):2087-2097. DOI: 10.1109/TSTE.2019.2950310

[22] Waewsak J, Ali S, Natee W, Kongruang C, Chancham C, Gagnon Y. Assessment of hybrid, firm renewable energy-based power plants: Application in the southernmost region of Thailand. Renewable and Sustainable Energy Reviews. 2020;**130**:109953. DOI: 10.1016/j.rser.2020.109953

[23] Liu J, Yin Y, Yan S. Research on clean energy power generation-energy storage-energy using virtual enterprise risk assessment based on fuzzy analytic hierarchy process in China. Journal of Cleaner Production. 2019;**236**:117471. DOI: 10.1016/j.jclepro.2019.06.302

[24] Jiang T, Li X, Chen H, Zhang R, Li G, Kou X, et al. Optimal energy storage siting and sizing to mitigate voltage deviation in distribution networks, in. IEEE Power Energy Society General Meeting (PESGM). 2019;**2019**:1-5. DOI: 10.1109/PESGM40551.2019.8973500

[25] Al-Madhlom Q, Al-Ansari N, Laue J, Nordell B, Hussain HM. Site selection of aquifer thermal energy storage systems in shallow groundwater conditions. Water. 2019;**11**(7):1393

[26] Solangi YA, Longsheng C, Shah SAA. Assessing and overcoming the renewable

energy barriers for sustainable development in Pakistan: An integrated ahp and fuzzy topsis approach. Renewable Energy. 2021;**173**:209-222

[27] Zanakis SH, Solomon A, Wishart N, Dublish S. Multi-attribute decision making: A simulation comparison of select methods. European Journal of Operational Research. 1998;**107**(3): 507-529. DOI: 10.1016/S0377-2217(97) 00147-1

[28] Daim T, Yates D, Peng Y, Jimenez B. Technology assessment for clean energy technologies: The case of the pacific northwest. Technology in Society. 2009; **31**(3):232-243. DOI: 10.1016/j.techsoc.2009.03.009

[29] Yuan X-C, Lyu Y-J, Wang B, Liu Q-H, Wu Q. China's energy transition strategy at the city level: The role of renewable energy. Journal of Cleaner Production. 2018;**205**:980-986. DOI: 10.1016/475j.jclepro.2018.09.162

[30] Luthra S, Kumar S, Garg D, Haleem A. Barriers to renewable/sustainable energy technologies adoption: Indian perspective. Renewable and Sustainable Energy Reviews. 2015; **41**:762-776. DOI: 10.1016/j.rser.2014.08.077

[31] Cowan K, Daim T, Anderson T. Exploring the impact of technology development and adoption for sustainable hydroelectric power and storage technologies in the pacific northwest united states. Energy. 2010; **35**:4771-4779. DOI: 10.1016/j.energy.2010.09.013

[32] Ben Ammar F, Hafsa IH, Hammami F. Analytic hierarchy process selection for batteries storage technologies. In: 2013 International Conference on Electrical Engineering and Software Applications. Hammamet, Tunsia. 2013. pp. 1-6. DOI: 10.1109/ICEESA.2013.6578374

[33] Walker SB, Mukherjee U, Fowler M, Elkamel A. Benchmarking and selection of power-to-gas utilizing electrolytic hydrogen as an energy storage alternative. International Journal of Hydrogen Energy. 2016;**41**:7717-7731. DOI: 10.1016/j.ijhydene.2015.09.008

[34] Ayodele T, Ogunjuyigbe A, Odigie O, Munda J. A multi-criteria gis based model for wind farm site selection using interval type-2 fuzzy analytic hierarchy process: The case study of Nigeria. Applied Energy. 2018;**228**: 1853-1869. DOI: 10.1016/j.apenergy.2018.07.051

[35] Wu B, Yip TL, Xie L, Wang Y. A fuzzy-madm based approach for site selection of offshore wind farm in busy waterways in China. Ocean Engineering. 2018;**168**:121-132. DOI: 10.1016/j.oceaneng.2018.08.065

[36] Ali S, Taweekun J, Techato K, Waewsak J, Gyawali S. Gisbased site suitability assessment for wind and solar farms in Songkhla, Thailand. Renewable Energy. 2019;**132**:1360-1372. DOI: 10.1016/j.renene.2018.09.035

[37] Uyan M. Gis-based solar farms site selection using analytic hierarchy process (ahp) in Karapinar region, Konya/Turkey. Renewable and Sustainable Energy Reviews. 2013;**28**: 11-17. DOI: 10.1016/j.rser.2013.07

[38] Wang Q, Han R, Huang Q, Hao J, Lv N, Li T, et al. Research on energy conservation and emissions reduction based on ahp-fuzzy synthetic evaluation model: A case study of tobacco enterprises. Journal of Cleaner Production. 2018;**201**:88-97. DOI: 10.1016/j.jclepro.2018.07.270

[39] How BS, Yeoh TT, Tan TK, Chong KH, Ganga D, Lam HL. Debottlenecking of sustainability performance for integrated biomass supply chain: P-graph approach. Journal of Cleaner Production. 2018;**193**:720-733. DOI: 10.1016/j.jclepro.2018.04.240.525

[40] Ma Y, Shi T, Zhang W, Hao Y, Huang J, Lin Y. Comprehensive policy evaluation of nev development in China, Japan, the United States, and Germany based on the ahp-ew model. Journal of Cleaner Production. 2019;**214**:389-402. DOI: 10.1016/j.jclepro.2018.12.119

[41] Saaty TL, Vargas LG. How to make a decision. Models, Methods, Concepts and Applications of the Analytic Hierarchy Process. Springer; 2012;**175**: 23-40. DOI: 10.1007/978-1- 4614-3597-6

[42] Luo X, Wang J, Dooner M, Clarke J. Overview of current development in electrical energy storage technologies and the application potential in power system operation. Applied Energy. 2015; **137**:511-536. DOI: 10.1016/j. apenergy.2014.09.081

[43] Ferreira JH, Tavares C. Desenvolvimento, implementação e simulação de um controlador para sistemas de armazenamento de energia com baterias [thesis]. Porto, Portugal: University of Porto; 2015

[44] Huff G. DOE Global Energy Storage Database. Albuquerque, NM (United States): Sandia National Lab; 2015

Chapter 12

Analytic Hierarchy Process (AHP) Applications in Watershed Management Plan, A Case Study of Sub-Watershed

Kartic Bera, Jatisankar Bandyopadhyay and Pabitra Banik

Abstract

Analytical Hierarchy Process (AHP) one of the Multi-Criteria Evaluation techniques is proposed by Saaty's-1980. More than one parameter is equally important for management, as these are very much interrelated. One of the methods is AHP, which is welcomed for supporting procedural justice that regards clearness and equality of decisions. However, the AHP method promises for procedural justice are partly grounded in it is supposed by numerical accuracy. By contrast, AHP method can contribute to the multiple criteria for procedural justice, which may explain AHP's continuing and growing popularity. The actual process of applying the decision rule is evaluation. In order to meet a specific objective, several criteria need to be evaluated. Such a procedure is termed as AHP. The integration of special information technology and multi-criteria methods facilitate the provision of a tool with great potential for obtaining watershed boundary or selecting sites for taking action plan. AHP provides an appropriate framework for the application of multi-criteria evaluation methods, whereas multi-criteria evaluation techniques add to GIS the means of performing trade-offs on contradicting objectives, giving weight-age to both multiple criteria as well as the knowledge of the decision maker.

Keywords: AHP, Dryland, water balance, surface water, prioritization, management plan

1. Introduction

According to Thomas L. Saaty [1] "We are all fundamentally decision makers. Everything we do consciously or unconsciously is the result of some decision". By the objectives, criteria, sub-criteria, alternatives multi-level hierarchical structure are used. Pair wise comparisons pertinent data set derived from pair wise comparison [2]. These pair wise comparisons are used to obtain the weights of rank of the decision criteria, and the relative performance measures of the alternatives in terms of each individual decision criterion. The mechanism improving is required if the comparisons are not perfectly consistent.

Analytical Hierarchy Process (AHP) one of the Multi-Criteria Evaluation techniques is proposed by Saaty's-1980. The AHP is increasingly used as decision support method along with multiple parameters. AHP is welcomed for supporting procedural justice that regards clearness and equality of decisions. The numerical basis of AHP is not as unequivocal as current 'AHP standard practice' suggests. By contrast, AHP can contribute to the multiple criteria for procedural justice, which may explain AHP's continuing and growing popularity. It is increasingly used as decision support method along with multiple parameters.

Many researches had been applied in several fields, for example, in engineering [3], industry [4], economics [5], environmental management, and water management [6–10].

Recently researchers have applied AHP to decision-making concerning water resource management [9, 11] focusing on criteria relating to social, economic [12], and environmental factors [13]. Thus, multi-criteria (or multi-parameters) decision support methods are widely applied in evaluation and strategic watershed planning and management, and infrastructure development. Multiple criteria analysis techniques have been used by water resource practitioners to select or to design alternatives in areas such as river basin planning and development, water resources development, land use management, groundwater/surface water allocation, watershed restoration and water resources quality [14, 15].

More than one parameter is equally importance for management [16], as this is very much interrelated. In the present study watershed management, 6 standard methods were used and studied briefly to calculate individual parameters of the study area. Final priority has been evaluated through AHP method [1]. This is useful for water resource management, with diverse malty parameter, for prioritization questions with diverse criteria or for allocation of scarce resources. However, AHP's promises for procedural justice are partly grounded in its supposed numerical accuracy.

Insufficient of hydrologic data, the logical alternative is to use for estimate the hydrologic characteristics of a watershed by using drainage morphometric parameters. These parameters can be precisely estimated in GIS environment. Extensively used techniques for estimating direct runoff depths from storm rainfall proposed by United States Department of Agriculture (USDA) Soil Conservation Service's (SCS) Curve Number (CN) method [17]. SYI model used for quantitative assessment of soil erosion which is basic aspect of watershed management, one can predict the amount of soil loss by using some empirical formulae. Also demographic stretcher information is required for best management practices on these areas and assessment of BMP implementation effectiveness on water amiability improvement through monitoring strategies. The final priority and action plan of each micro-watershed has been taken by Analytic Hierarchy Process (AHP) [18].

2. Case study area

Dwarakeswar (also known as Dhalkishor) river originates from the Tilaboni hill near Bangalia Railway Station in Puruliya district of West Bengal state, flows easterward through pedimental landscape and enters into dissected Bankura, then enters in lateritic upland and further downstream into the Gangetic Alluvial Terrain debouching in the Rupnarayan river [19]. Beko is one of the watersheds of Dwarakeswar river in Bankura and Puruliya district. Upper Beko watershed (2A2C8) contains 7 sub-watersheds i.e. Beko, Dangra, Kansachara, Arkasa, Dudhbhariya, Darubhanga and Futuari (**Figure 1**). This study was undertaken in sub-watersheds namely Dangra (**Figure 1**).

3. Dangra river

Left hand tributary of Dwarkeswar river flowing through Kasipur, Raghunathpur-I, & Santuri block of Puruliya district and Chhatna block in Bankura district. Dangra sub-watershed contain by proposed 29 micro-watersheds (**Figure 2**). The area of Dangra sub-watershed is 210.16 km^2 and elevation varies between 90 m to 250 m above mean sea level (MSL).

The area under study constitutes part of a semi-arid region with distinct characteristics in terms of the following elements i.e. Relief, Soil, Geology, Climate, Drainage, and Slope, Land use / Land cover and Groundwater availability.

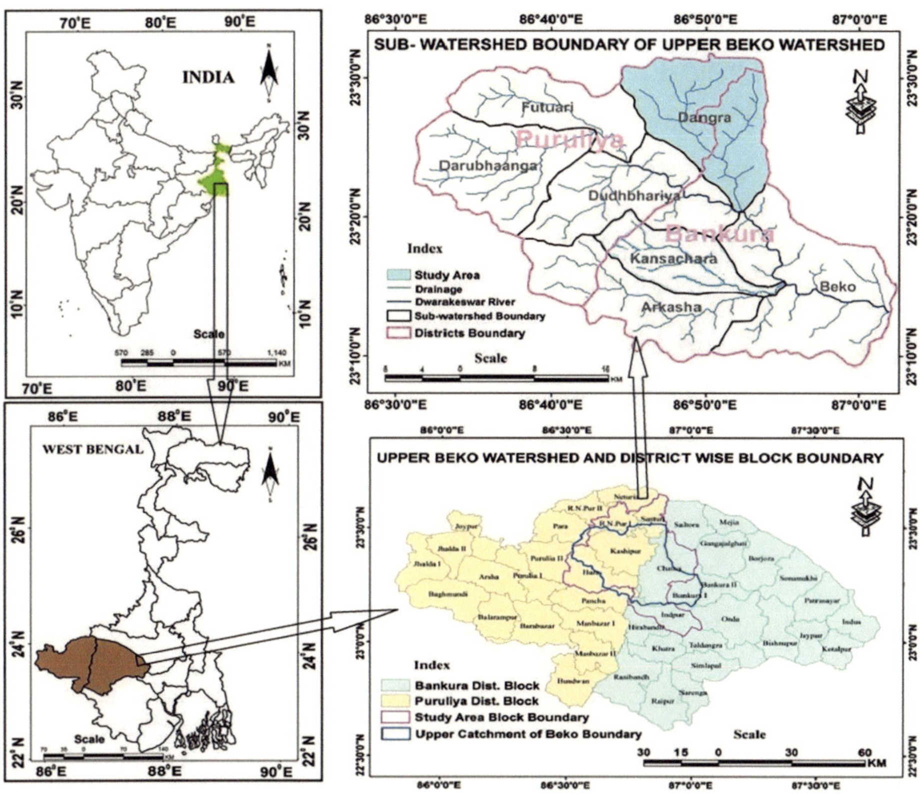

Figure 1.
Location of study area.

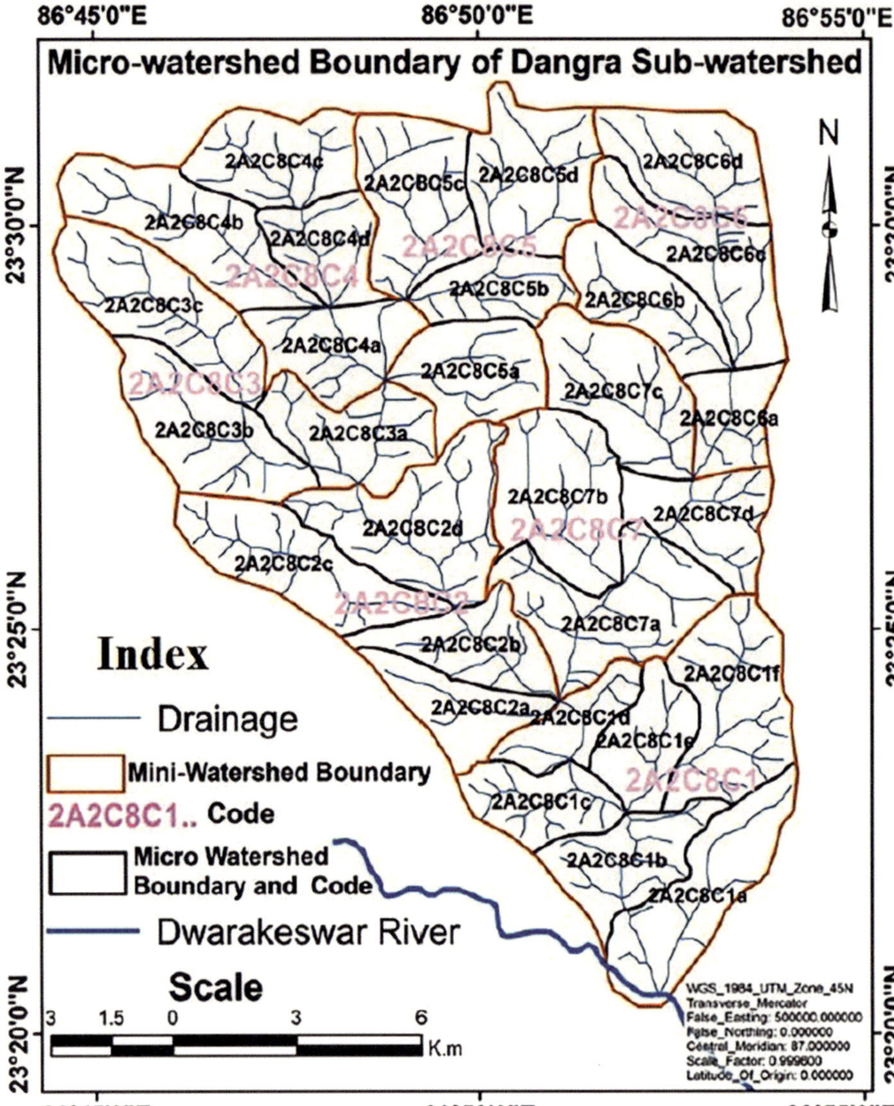

Figure 2.
Study area watershed.

4. Methodology

Watershed management is described by the different natural parameters for tacking appropriate action plan or implementation. Watershed management is a complex attribute of watershed that has a direct effect on micro watershed. Most micro watershed parameters are determined by the interaction of several characteristics and measurable natural attributes of the region [19, 20]. In this study, selected numbers of parameters have been selected for evaluation the micro watershed. Once the problem has been recognized, the set of criteria for assessment needs to be designated.

Assigned value	Definition	Explanation
1	Parameters are of equal importance	Two parameters contribute equally to the objective
3	Parameter j is of weak importance compared to parameter i	Experience and Judgment slightly favor parameter i over j
5	Essential or strong importance of parameter i compared to j	Experience and Judgment strongly favor parameter i over j
7	Demonstrated importance	Criteria i is strongly favored over j and its dominance is demonstrated in practice
9	Absolute importance	The evidence favoring parameter i over j to the highest possible order of affirmation
2,4,6 & 8	Intermediate values between two adjacent judgment	Judgment is not precise enough to assign values of 1,3,5,7,and 9

Table 1.
Scale for pair-wise comparison.

Since the evaluation criteria are related to geographical entities and the interlinking between them, they can be approximately represented.

The actual process of applying the decision rule is evaluation. In order to meet a specific objective several criteria need to be evaluated. Such a procedure is termed as Multi-Criteria Evaluation. The integration of special information technology and multi-criteria methods facilitate the provision of a tool with great potential for obtaining micro watershed boundary or selecting sites for taking action plan [19]. AHP provides and appropriate framework for the application of multi-criteria evaluation methods, whereas multi-criteria evaluation techniques add to GIS the means of performing trade-offs on contradicting objectives, giving weightage to both multiple criteria as well as the knowledge of the decision maker.

The multi-criteria decision analysis method is widely used for pair-wise comparison techniques. Analytic Hierarchy Process (AHP) is developed by Saaty [1] for decision making process. The pair-wise comparison of related parameters results into the 'importance matrix' which is based on a scale of importance intensities. **Table 1** is elaborately discus about scale of importance. The importance matrix can then be analyzed by various methods-"Eigen-Vector method" or "Least Square" method, to arrive at the weightages of each parameter in the matrix. However, in the present study a ratio (reciprocal matrix) is constructed, where each factor or criteria is compared with the other criteria, relative to its importance on a scale from 1 to 9.

Weights are calculated by normalizing the eigen-vector associated with the maximum eigen-value of the matrix. This involves the following operation:

- Computation of sum of values in each column of the Pair-wise comparison matrix;
- Normalization of the matrix by dividing each element by its columan total;
- Computation of mean of the elements in each row of the normalized matrix.

Then the consistency ratio (CR) is computed to check the consistency of comparisons by using the flowing formulas:

$$CR = \frac{\text{Consistancy Index}\,(CI)}{\text{Random Inconsistency Index}\,(RI)} \quad (1)$$

$$CI = \frac{\text{ë max} - n}{n - 1} \quad (2)$$

Where.

λ max – Principal Eigen – value = sum of products between each elements of the priority vector and column total.

n = number of comparisons/criteria.

If the consistency ratio (CR) > 0.10, then some pair-wise values need to be reconsidered and the process is repeated till the desired value of CR < 0.10 is reached (**Table 2** and **Figure 3**).

Number of criteria (n)	Random Inconsistency Indices (RI)
1	0.00
2	0.00
3	0.58
4	0.90
5	1.12
6	**1.24**
7	1.32
8	1.41
9	1.45
10	1.49

Table 2.
Random Inconsistency Indices (RI).

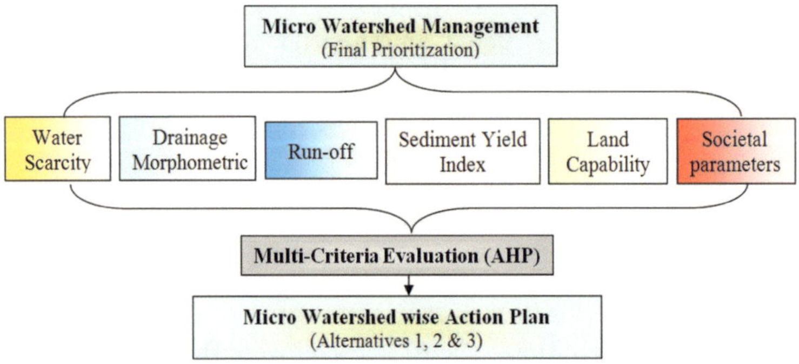

Figure 3.
Flow chart of the work.

5. Result and discussion

The methodology described above has been implemented for the Dangra sub-watersheds to determine the suitable of land for micro-watershed management and taking action plan [21]. The prioritization at micro watershed level for the sub-watersheds based on the following parameters; Water scarcity, Sediment Yield Index, Land Capability, Morphomatic parameters, Societal parameters, and Run-off (**Tables 3** and **4**).

Calculation of Consistency Ratio (CR):

$$\text{ë max} = (1.928\times 46.07632855)+(4.923\times 24.29530124) \\ +(10.033\times 12.76595183)+(15.7\times 8.592603084) \\ +(19.5\times 5.026182866)+(27\times 3.243632435)=6.57014234 \quad (3)$$

$$CI=(6.57014234-6)/(6-1)=0.114028468 \quad (4)$$

$$CR=\frac{0.114028468}{1.24}(6\,\text{Parameters,Value of }RI=1.24) \\ =0.0919584\left[\text{Consistency is acceptable, } as\ CR<0.10\right] \quad (5)$$

6. Conclusion

As per the analysis in **Table 5** of the total 29 micro-watershed in the sub-watershed eighteen micro-watersheds are fall under fast management stage twelve micro-watersheds, in second management stage seven micro-watershed and four under third management stage, whereas immediate action not required for development based on sub-watersheds (**Figure 4**).

The multi-criteria evaluations of physical parameters of the sub-watersheds indicate that the area is facing a great water resources problem. Thematic information of scarcity zone, morphometric characters, surface runoff, sediment yield, land capability and socio-economy with demography were generated in the GIS environment

Factors	Scarcity	SYI	Land capability	Morphometric	Societal	Run-off
Scarcity	1	3	5	7	7	9
SYI	1/3	1	3	5	4	7
Land capability	1/5	1/3	1	2	5	3
Morphometric	1/7	1/5	1/2	1	2	5
Societal	1/7	1/4	1/5	1/2	1	2
Run-off	1/9	1/7	1/3	1/5	1/2	1
Sum	1.928	4.923	10.033	15.7	19.5	27

Table 3.
Pair-wise comparison of factors for action Plan.

Factors	Scarcity	SYI	Land capability	Morphometric	Societal	Run-off	Sum	Priority vector	% of Weight
Scarcity	0.51867	0.60930	0.4983554	0.4458598	0.3589743	0.333333	2.764579	0.4607632	46.076329
SYI	0.17271	0.20312	0.2990132	0.3184713	0.2051282	0.259259	1.457718	0.2429530	24.295301
Land capability	0.10373	0.0676	0.0996710	0.1273885	0.2564102	0.111111	0.765957	0.1276595	12.765952
Morphometric	0.07365	0.0406	0.0498355	0.0636942	0.1025641	0.185185	0.515556	0.0859260	8.5926031
Societal	0.07365	0.0507	0.0199342	0.0318471	0.0512820	0.074074	0.301570	0.0502618	5.0261829
Run-off	0.05757	0.0284	0.0331904	0.0127388	0.0256410	0.037037	0.194617	0.0324363	3.2436324

Table 4.
Normalized matrix with results.

DOI: http://dx.doi.org/10.5772/intechopen.1001456

Factors	Scarcity	SYI	Land capability	Morphometric	Societal	Run-off	Total	Action Plan
MWC								
2A2C8C1a	46.07633	24.2953	12.76595183	8.592603084	10.05237	6.487265	108.2698	1
2A2C8C1b	46.07633	48.5906	12.76595183	8.592603084	10.05237	3.243632	129.3215	1
2A2C8C1c	138.229	48.5906	12.76595183	8.592603084	10.05237	3.243632	221.4741	2
2A2C8C1d	46.07633	48.5906	12.76595183	17.18520617	15.07855	3.243632	142.9403	1
2A2C8C1e	46.07633	48.5906	12.76595183	25.77780925	10.05237	6.487265	149.7503	1
2A2C8C1f	46.07633	24.2953	12.76595183	25.77780925	10.05237	3.243632	122.2114	1
2A2C8C2a	138.229	48.5906	12.76595183	17.18520617	10.05237	6.487265	233.3104	3
2A2C8C2b	92.15266	24.2953	12.76595183	17.18520617	10.05237	6.487265	162.9387	1
2A2C8C2c	138.229	48.5906	12.76595183	8.592603084	10.05237	9.730897	227.9614	3
2A2C8C2d	92.15266	24.2953	12.76595183	8.592603084	10.05237	3.243632	151.1025	1
2A2C8C3a	92.15266	24.2953	25.53190366	17.18520617	10.05237	3.243632	172.4611	2
2A2C8C3b	138.229	24.2953	12.76595183	25.77780925	10.05237	6.487265	217.6077	2
2A2C8C3c	92.15266	24.2953	25.53190366	25.77780925	10.05237	9.730897	187.5409	2
2A2C8C4a	92.15266	24.2953	25.53190366	17.18520617	5.026183	9.730897	173.9221	2
2A2C8C4b	92.15266	72.8859	25.53190366	17.18520617	10.05237	9.730897	227.5389	3
2A2C8C4c	46.07633	24.2953	38.29785549	17.18520617	5.026183	9.730897	140.6118	1
2A2C8C4d	46.07633	24.2953	25.53190366	17.18520617	10.05237	9.730897	132.872	1
2A2C8C5a	46.07633	24.2953	12.76595183	25.77780925	15.07855	6.487265	130.4812	1
2A2C8C5b	46.07633	48.5906	25.53190366	17.18520617	10.05237	9.730897	157.1673	1
2A2C8C5c	46.07633	24.2953	38.29785549	25.77780925	5.026183	9.730897	149.2044	1
2A2C8C5d	46.07633	24.2953	38.29785549	25.77780925	5.026183	9.730897	149.2044	1
2A2C8C6a	46.07633	72.8859	12.76595183	17.18520617	10.05237	9.730897	168.6967	2

Factors	Scarcity	SYI	Land capability	Morphometric	Societal	Run-off	Total	Action Plan
MWC								
2A2C8C6b	138.229	72.8859	25.53190366	17.18520617	10.05237	9.730897	273.6153	3
2A2C8C6c	138.229	24.2953	25.53190366	8.592603084	5.026183	9.730897	211.4059	2
2A2C8C6d	46.07633	48.5906	38.29785549	8.592603084	5.026183	9.730897	156.3145	1
2A2C8C7a	46.07633	24.2953	12.76595183	25.77780925	10.05237	3.243632	122.2114	1
2A2C8C7b	46.07633	24.2953	12.76595183	17.18520617	10.05237	3.243632	113.6188	1
2A2C8C7c	46.07633	48.5906	12.76595183	17.18520617	5.026183	6.487265	136.1315	1
2A2C8C7d	46.07633	48.5906	12.76595183	17.18520617	10.05237	6.487265	141.1577	1

Table 5.
AHP based Sub-Watershed Final Prioritization.

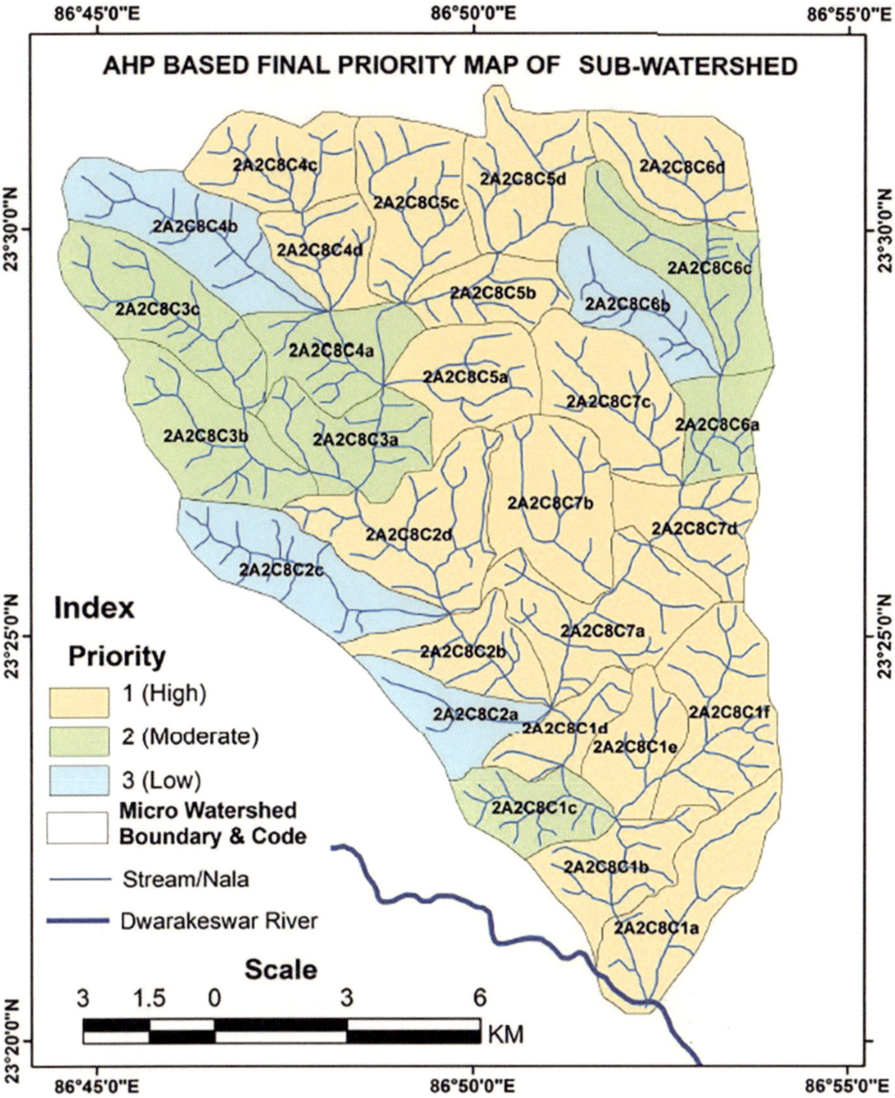

Figure 4.
AHP based final priority.

using remote sensing data and field data. Therefore, all final prioritized value is assigned weightage to obtain normalized weightage by using analytical hierarchy process. Then all the parameters are integrated in the GIS environment to decide the soil conservation measures at the appropriate site in the watershed.

AHP provides an appropriate framework for the application of multi-criteria evaluation methods, whereas multi-criteria evaluation techniques add to GIS the means of performing trade-offs on contradicting objectives, giving weight-age to both multiple criteria as well as the knowledge of the decision maker. On the whole, this study demonstrated that the remote sensing, GIS and AHP techniques offer a useful integrated tool for the deciding micro watershed management plan with societal perspective towards development and planning.

Author details

Kartic Bera[1*], Jatisankar Bandyopadhyay[2] and Pabitra Banik[3]

1 Geography in the Department of Pure and Applied Science, Midnapore City College, Midnapore, West Bengal, India

2 Department of Remote Sensing and GIS, Vidyasagar University, Midnapore, West Bengal, India

3 Agricultural and Ecological Research Unit, Indian Statistical Institute, Kolkata, West Bengal, India

*Address all correspondence to: 1kbrsgis@gmail.com

References

[1] Saaty TL. The analytic hierarchy process: Planning, priority setting. Resource Allocation. 2008;**1**:287

[2] Brunelli M. Introduction to the analytic hierarchy process. Springer briefs in. Operations Research. 2015;**2015**:83. DOI: 10.1007/978-3-319-12502-2

[3] Rezakhani P. Fuzzy MCDM model for risk factor selection in construction projects. Engineering Journal. 2012;**16**(5):81-93. DOI: 10.4186/ej.2012.16.5.79

[4] Dweiri F, Kumar S, Khan SA, Jain V. Designing an integrated AHP based decision support system for supplier selection in automotive industry. Expert Systems with Applications. 2016;**62**:273-283. DOI: 10.1016/j.eswa.2016.06.030

[5] Riahi A, Moharrampour M. Evaluation of strategic management in business with AHP case study: PARS house appliance. Procedia Economics and Finance. 2016;**26**:10-21. DOI: 10.1016/S2212-5671(16)30011-9

[6] Bosch D, Pease J, Wolfe ML, Zobel C, Osorio J, Cobb TD, et al. Community DECISIONS: Stakeholder focused watershed planning. Journal of Environmental Management. 2012;**112**:226-232. DOI: 10.1016/j.jenvman.2012.07.031

[7] Gallego-Ayala J, Juízo D. Integrating stakeholders' references into water resources management planning in the Incomati River Basin. Water Resource Management. 2014;**28**:527-540. DOI: 10.1007/s11269-013-0500-3

[8] Montazar A, Zadbagher E. An analytical hierarchy model for assessing global water productivity of irrigation networks in Iran. Water Resource Management. 2010;**24**:2817-2832. DOI: 10.1007/s11269-010-9581-4

[9] Srdjevic B, Medeiros YDP. Fuzzy AHP assessment of water management plans. Water Resources Management. 2008a;**22**:877-894. DOI: 10.1007/s11269-007-9197-5

[10] Shabbir R, Ahmad SS. Water resource vulnerability assessment in Rawalpindi and Islamabad, Pakistan using Analytic Hierarchy Orocess (AHP). Journal of King Saud University—Science. 2016;**28**(4):293-299. DOI: 10.1016/j.jksus.2015.09.007

[11] Anagnostopoulos KP, Petalas C, Pisinaras V. Water resources planning using the AHP and PROMETHEE multicriteria methods: The case of Nestos River-Greece. In: Presented at the 7th Balkan Conference on Operational Research, May 25-28, 2005. Constanta, Romania; 2005

[12] Tangtham N. Status of socio-economic-institutional research and identification of needed researches for watershed management in Thailand. Kasetsart Journal of Social Sciences. 1986;**7**(2):217-226. Retrieved from: https://so04.tci-thaijo.org/index.php/kjss/article/view/243773

[13] Thungngern TS, Wijitkosum S. Analytic hierarchy process for stakeholder participation in integrated water resources management. Engineering Journal. 2017;**21**(7):87-103

[14] Mirchi A, Watkins D, Madani K. Modeling for watershed planning, management, and decision making. In: Vaughn JC, editor.

Watersheds: Management. Restoration and Environmental Impact. Environmental Science. Engineering and Technology Series. 2010. e-book. ISBN: 978-1-61209-295-9. Available from: http://www.kysq.org/docs/Mirchi_Watershed.pdf

[15] Ceballos-Silva A, Lopez-Blanco J. Delineation of suitable areas for crops using a multi-criteria evaluation approach and land use/ land cover mapping: A case study in Central Mexico. Agricultural Systems. 2003;**77**(2):117-136. DOI: 10.1016/S0308-521X(02)00103-8

[16] World Bank Report. Thailand Environment Monitor: Integrated Water Resources Management—A Way Forward [Online]. 2017. Available from: http://documents.worldbank.org/curated/en/367151468303847751/Thailand-environment-monitor-integrated-water-resources-management-a-way-forward

[17] USDA Soil Conservation Service. National Engineering Handbook, Section 4, Hydrology. Washington, DC, USA: USDA-SCS; 1985

[18] Malczewski J. GIS based multi-criteria decision analysis: A survey of the literature. International Journal of Geographical Information Science. 2006;**20**:703-726. DOI: 10.1080/13658810600661508

[19] Bera K, Banik P. Multi-criteria decision analysis (MCDA) for surface water management plan, a case study of Kansachara sub-watershed, West Bengal, India. Water Supply. 2019, 2019;**19**(7):2156-2162. DOI: 10.2166/ws.2019.096

[20] Chakhar S, Mousseau V. GIS based multi-criteria spatial modeling generic framework. International Journal of Geographic Information Science. 2008;**22**:1159-1196. DOI: 10.1080/13658810801949827

[21] Kumar B, Kumar U. Micro watershed characterization and prioritization using geomatics technology for natural resources management. International Journal of Geomatics and Geosciences. 2011;**1**(4):789-802